The Biased Mind

Jérôme Boutang · Michel De Lara

The Biased Mind

How Evolution Shaped Our Psychology
Including Anecdotes and Tips
for Making Sound Decisions

Jérôme Boutang
Paris, France

Michel De Lara
Paris, France

ISBN 978-3-319-16518-9   ISBN 978-3-319-16519-6 (eBook)
DOI 10.1007/978-3-319-16519-6

Library of Congress Control Number: 2015951247

Springer Cham Heidelberg New York Dordrecht London
© Springer International Publishing Switzerland 2016

This work is subject to copyright. All rights are reserved by the Publisher, whether the whole or part of the material is concerned, specifically the rights of translation, reprinting, reuse of illustrations, recitation, broadcasting, reproduction on microfilms or in any other physical way, and transmission or information storage and retrieval, electronic adaptation, computer software, or by similar or dissimilar methodology now known or hereafter developed.

The use of general descriptive names, registered names, trademarks, service marks, etc. in this publication does not imply, even in the absence of a specific statement, that such names are exempt from the relevant protective laws and regulations and therefore free for general use.

The publisher, the authors and the editors are safe to assume that the advice and information in this book are believed to be true and accurate at the date of publication. Neither the publisher nor the authors or the editors give a warranty, express or implied, with respect to the material contained herein or for any errors or omissions that may have been made.

Printed on acid-free paper

Springer International Publishing Switzerland is part of Springer Science+Business Media (www.springer.com)

*"At fifteen, I became an evolutionist, and it all became clear.
We came from mud.
And after 3.8 billion years of evolution, at our core is still mud.
Nobody can be a divorce lawyer and doubt that."*

*Gavin (Danny DeVito) in
The War of the Roses (directed by Danny DeVito, 1989)*

Preface

In 2009, Michel and Jérôme embarked on a research project on risk perception with the Paris School of Economics.

Michel is an academic, professor, and researcher at the French institution Ecole des Ponts ParisTech. He develops mathematical methods for the sustainable management of natural resources—like fisheries, epidemics, and renewable energies—which involves among other things investigating the mathematical and economic aspects of risks.

Jerome had been a marketing professional for twenty years when, as an international consultant, he came across the issues of sustainability and strategy in the food and wine industry. He soon widened his expertise to people's perception of ecological threats, such as climate change and air pollution, which seemed to be the core motivation for moving towards sustainability, and to be the source of a problematic gap between corporations and public decision-makers, and their audience.

The project at the Paris School of Economics dealt with human perceptions and motivations. What makes people "tick"? How does each of us, family, friends, colleagues, movie characters, politicians, etc., form beliefs, make choices, and act?

Reading dozens of books and hundreds of articles, we had a Eureka experience.

Like Danny DeVito's character, suddenly *"it all became clear"*: catching how evolution shaped our psychology helps to figure out how we connect with others, forge beliefs, and make suitable choices.

This is how *The Biased Mind* project was ignited.

We turned back to the material we had accumulated in economics, cognitive science, neuroscience, and evolutionary psychology. As we grew more aware of our mind biases, we decided to package the fascinating ideas we had come across in palatable snacks, full of visuals and anecdotes, our mind's mother tongue.

We soon realized that, by better sizing the adaptive biases of our mind, we could better appraise, decide, and communicate. So we have bottled useful tips for you, whether for job interviews, weighing the pros and cons in order to reach a decision, appraising marital happiness, becoming a wine tasting expert, or framing messages.

Reading *The Biased Mind* should give you a better feel for, and a better ability to act on your beliefs, choices, and relationships. You should also be able to answer intriguing questions like:

> Why is an upside-down red triangle such a powerful warning sign?
> How does one produce a good alibi?
> What makes the number 7 so special?
> Will your recent marriage last?
> Why do the French eat snails not slugs?
> Why is our brain tuned towards paranoia?

These questions have to do with our mind's limitations and biases. On a more mundane level, we tend to have strong first impressions and find it hard to change our views; we often avert regret and waver between gut instinct and reason. At other times, we may feel influenced as if just the right chords have been struck in our mind.

What are mental biases? Where do they come from? You may be surprised to discover that the Pleistocene hunter-gatherer still lurks in our minds and influences our daily assessments and decisions. By becoming aware of the various aspects of human bias and applying our helpful tips, you should be better equipped to understand yourself and others, to interact in different social contexts, reach better judgements, make better decisions, avoid manipulation, and communicate more efficiently.

This is what *The Biased Mind* is about.

Key

In the following, you will come across three symbols or pictograms.

The symbol 👁 signals a helpful tip dealing with a specific bias.

The symbol 📖 introduces a small dose of theory.

And finally, the pictogram ⧖ indicates an amusing test.

At the end of *The Biased Mind* you will find a list of all the anecdotes and tips, should you wish to choose "à la carte" from our mouth-watering menu.

Acknowledgements and Other References

Follow us on ourwebsite
www.thebiasedmind.com
(From the authors of *The Biased Mind*).

Our special thanks go to
Ecole des Ponts ParisTech and the International Technical Centre for Studies on Air Pollution (CITEPA) for the conditions of work they provide,

Ecole des Ponts ParisTech students for two of the pictures and for some nice bibliographical work,

Jean-Marc Tallon, Paris School of Economics (PSE), who headed the research project AXA–PSE entitled "The economics and psychology of risk taking, impatience and financial decisions: confronting survey, experimental and insurance data", and all the team,

Angela Lahee from Springer Publishing who has been extremely supportive from the start and gave us invaluable advice on the overall orientation; we would especially like to thank Stephen Lyle for his careful, always instructive and precise editing and recommendations; we are also most grateful to Deirdre Nuttall, Adverbage Ltd, for her assistance in clarifying, organising, and editing material during the preparation of this book; also David Woodruff for his most helpful advice, Mark Tuddenham for his checking and editing, Maria Gallardo for two pictures in Pinamar, Jean-Pierre Chang for a nice picture of an eye, and the Phelps family for their checking.

Contents

1. Introduction 1
2. Embarking On the Mind Tour 5
3. Better Be Paranoid to Survive 19
4. We Like Things the Way They Are 49
5. Our Detective Mind Grasps Clues and Narrates 65
6. Images Call More to Mind Than Words and Numbers 89
7. How to Balance Pros and Cons, and Other Helpful Hints 123
8. I Frame, You're Framed 143
9. Epilogue: Does It Really Pay to Weigh Up Our Biases? 171

Detailed Contents 173
Bibliography 179

1
Introduction

We invite you on an entertaining journey inside the mind maze.

We have all learned, from Darwin's theory of evolution, that the form and function of the various organs our body is made up of—lungs, liver, heart, stomach, etc.—result from millennia of evolution through natural selection. In biology, an organ is a collection of tissues dedicated to serve a common function. The heart pumps and channels the blood through the body. Lungs allow the transfer of oxygen to the blood. Bones support the body. Muscles enable movement. Stomach, glands, and intestine are part of the food processing function. Immune systems protect from disease. The nervous system gathers and transmits information to the brain, and the brain itself is an organ.

In fact, the brain is a collection of mental organs, also called "modules", each with a different function. Just as body organs specialize in solving specific problems—the liver detoxifying poisons, sexual organs ensuring reproduction, etc.—so there are "mental organs" for face recognition, mate finding, inferring people's behavior, and so on. One can view the brain as a bundle of dedicated chips whose operations generate behavior. As a consequence, speaking of "a" brain function is delicate. As neuroscientist Joseph LeDoux[1], Professor at New York University, puts it: *"There is no equation by which the combination of functions of all the different systems mixed together equals an additional function called brain function."*

Charles Darwin expressed the fascinating point of view that our behaviors, attitudes, and cognitive processes have also been shaped by natural selection. He even dared to claim that *"He who understands baboon would do more toward metaphysics than Locke"*. The cornerstone book "Sociobiology. The New Synthesis"[2] by Edward O. Wilson, biologist and University Research Professor

---

[1] Ledoux, J., 1998
[2] Wilson, E. O., 2000

Emeritus at Harvard, synthesized a vast literature linking behaviour with natural selection. Viewing our mind from such an angle opens a rich research program.

Anthropologist John Tooby and psychologist Leda Cosmides, of the Center of Evolutionary Psychology at the University of California in Santa Barbara, were pioneers in the field of evolutionary psychology, which they define as "*an approach to psychology, in which knowledge and principles from evolutionary biology are put to use in research on the structure of the human mind*"[3].

Evolutionary psychology adopts the perspective that the mind is a collection of "*mental organs*" or "*modules*", fashioned during our common past as hunter-gatherers. That would be from circa 200 000 BC to circa 10 000 BC, if we consider Homo sapiens (our species); or even from 2 million years BC to 10 000 BC, if we take into consideration the emergence of the first Homo, our ancient ancestors, who gave birth to Homo sapiens (the modern man).

This "*new science of the mind*",[4] following the words of David Buss, Professor of Psychology at the University of Texas, formulates hypotheses and (refutable) empirical tests. A constellation of psychological observations—ranging from distorted perceptions to male-female relationships—can be explained by a limited array of biologically founded principles. Like Danny DeVito's character in The War of the Roses, we think that the "*adaptive*" point of view shines an insightful light on our psychology. By adopting the adaptive point of view with regard to the way people form beliefs, make choices, and act, many things suddenly become clear. As Cosmides and Tooby put it, "*Evolutionary psychology … illuminates the adaptations that constitute the machinery of behavior*".[5]

We have found it enlightening and rewarding to examine human psychology and biases—in the way we perceive, assess, and decide—as adaptive responses to long-lasting environmental and social pressures.

We have chosen the best stories to show the biased mind at work, selecting them from a vast array of literature spanning the fields of economics, anthropology, mathematics, neuroscience, psychology, sociology, ergonomics, marketing, and communication. We have ploughed through dozens of academic articles and science books so that you won't have to. Or maybe this will encourage you to open a few on your own!

Our minds love anecdotes and images, and it's been quite a job to dig out the best of these to put across our current understanding of biases. This fascinating tour of the mind begins in the remote past, when our hunter-gatherer ancestors lived in the savannah. We cannot experience their lives, but we share

---

[3] www.cep.ucsb.edu
[4] Buss, D., 2014
[5] Barkow, J., Cosmides, L. and Tooby, J., 1992

the same brain, a fantastic and complex organ that evolved to solve the many problems raised by survival and reproduction. Our brain is brimming with devices tailor-made to solve problems in environments that would rarely if ever concern us today. And many of our mental biases highlight the remoteness of those distant environments. We shall illustrate this with a handful of intriguing examples of adaptation mechanisms, some still relevant to modern day experience, but many not.

The term "bias" comes from Old Provençal (Occitan) "*biais*", meaning "*bend*", "*detour*", "*oblique turn*". Nowadays, when we speak of a "mental bias", we allude to a systematic tendency of our mind, and sometimes an irrational preference. Consider an analogy with games of chance. A bias at roulette is a systematic tendency for some numbers to come out slightly more often than others. If discovered, such a discrepancy could be used to the player's advantage, at the expense of the casino. This motivates casinos to devote time and energy to removing any mechanical biases. In his book *The Theory of Gambling and Statistical Logic*[6], Richard Arnold Epstein relates various tedious experiments in which a coin is tossed or dice rolled to detect possible biases. In 1894, English biologist Walter Frank Raphael Weldon reported the results of rolling a set of 12 dice 26,306 times. The frequency of five or six was 0.3377, when theoretically it should have been 0.3333, were the dice unbiased. The story is that Weldon's experiments boosted the development of statistical tests. Statistics show that a discrepancy as great as that between the empirical 0.3377 and the theoretical 0.3333 is vanishingly unlikely for such a large sample, so that difference does indeed reveal a significant bias. The bias is due to the fact that the five and six faces contain less matter than the opposite ones, carrying the numbers one or two, because of the hollowed-out pips in low quality dice. By gravity alone, the dice fall more often on the heavier faces, thus displaying five or six more often!

In the same vein, we can understand mental bias as a symptom of built-in adaptive mechanisms designed to maximize survival and reproduction. Examples of mental biases include the perception that losses loom larger than gains, or our mental urge to covet fatty foods, with the (undesirable) consequence of accumulating excessive body reserves, even though food is plentiful all year round in our affluent Western societies. A striking example of bias (to be expanded upon later) is our tendency, when an object is thrown at us, to judge it as coming closer than it really does. As Steven Pinker, Professor in the Department of Psychology at Harvard University has put it: "*our brains were shaped for fitness, not for truth. Sometimes the truth is adaptive, but sometimes it is not.*"[7]

---
[6] Epstein, R. A., 1977
[7] Pinker, S., 1997

The Biased Mind is everybody's mind, a perfectly sane and fit collection of mental organs that displays adaptive features.

Our journey will go on to explore our biases, from the way we perceive risk in various situations and the way we sample the world around us—extracting some clues to fill blanks but ignoring others, crafting scripts and stories—to the way we tend to visualize as much as we actually think.

We have illustrated *The Biased Mind* with a cocktail of examples from cinema, TV series, literature, sports, business, and daily life, and a small dose of our own work.

We will pave the way with useful tips that may help you to make better judgements and decisions, avoid manipulation, and communicate more efficiently.

2
Embarking On the Mind Tour

One of the authors heard this story during a scientific conference on ecology in Montpellier in 2010:

> Two little dinosaurs are running as fast as they can, chased by a large T. Rex.
> They are both exhausted and one says to the other:
> "Why bother running fast? We are stupid, it's hopeless,
> there's no way we can outrun a T. Rex."
> The other answers: "I'm not trying to run faster than the T. Rex,
> I'm trying to run faster than you!"

Sometime later, the same author heard a similar joke on the radio, telling the story of two travellers sitting quietly in a forest having lunch, when they see a bear coming. They quickly try to put their shoes on. One traveler says to the other: "Why bother?" You can guess what the other replied![1]

This is how evolution shaped our strategies for avoiding danger.

The corollary, in both cases, is that the faster of the two runners had a small advantage over the other one, and survived. This running advantage was transmitted to his or her descendants as a specific "predator avoidance module". Darwin named this process *"descent with modification"*. We now use the more appealing term *"evolution"*, as coined by his contemporary Herbert Spencer.

The theory of evolution has filled thousands of serious books and scientific papers; this is it in a nutshell. A full discussion goes beyond the scope of *The Biased Mind*[2].

On our journey in search of the adaptive biases, we shall encounter Ulysses and the Sirens and other long-suffering heroes, along with Dr Jekyll and Mr Hyde and the experience of multiple selves, not to mention the seven dwarves and other instances of the magical number seven. We shall eventually step ashore with a host of inherited problem-solving devices that can be put to

---
[1] You may not believe it, but three years later, he heard the same joke, but this time with a lion instead of a bear, during a conference on mathematics in Rio de Janeiro!
[2] Darwin, R.,1859; Dawkins, R., 1976; Barkow, J.H., Cosmides, L. and Tooby, J., 1992; Pinker, S. 1997.

work like a Swiss army knife to cope with the plethora of choices that face us on a daily basis.

Throughout our journey into the biased mind, two ideas in particular— that the brain is a partly outdated survival tool kit and that there are limitations on its capacity– will prove to be major recurring themes.

Who's the Boss?

Who is sovereign, who is in charge; the self who sets the alarm clock to rise early, or the self who shuts it down the next morning and goes back to sleep?[3]

This question, raised by economists George Loewenstein, Professor of Economics at Carnegie Mellon University, and Richard Thaler, Professor of Behavioral Science and Economics at the University at Chicago's Booth School of Business, will be the starting point in our quest to unmask the biased mind. At different times in our lives, we have all experienced the feeling that there may be, not just two, but several selves within us. We need only recall our dietary commitments on New Year's day and the manner in which they were gloriously ignored as the year got underway.

One brain—two minds?...asks Michael Gazzaniga, Professor of Psychology at the University of California, Santa Barbara, in his 1972 paper in *American Scientist*.... So let's explore this possibility of multiple selves.

Making Virtuous Choices

Future. That period of time in which our affairs prosper, our friends are true and our happiness is assured.
Ambrose Bierce.

Some of the virtuous choices that people make may involve a lack of empathy for the future self who will have to live with that choice.

In an elegant demonstration of this phenomenon, Daniel Read, George Loewenstein, and Shobana Kalyanaraman[4] provided experimental participants with coupons that allowed them to rent several films for free. There were two types of film: those that were edifying or "highbrow" (such as *Schindler's List*) while others were lowbrow and fun (such as *Sleepless in Seattle*). The films were available either for the same evening or for the next day. Subjects

---
[3] Loewenstein, G. F. and Thaler, R. H., 1989
[4] Read, D., Loewenstein, G. H. and Kalyanaraman, S., 1999

tended to select lowbrow movies for viewing tonight and highbrow movies for tomorrow.

The desire to improve one's mind is apparently more pressing when choosing a movie for later, whereas the desire to relax is more urgent when choosing for the very near future.

The Dumbledore Pact

Picture: Ulysses and the Siren- 1891, John William Waterhouse

On his journey home to Ithaca, Ulysses had to skirt the perilous land of the Sirens. Advised by the witch-goddess Circe to avoid them and their charming songs, Ulysses told his men[5]:

Take me and bind me to the crosspiece half way up the mast; bind me as I stand upright, with a bond so fast that I cannot possibly break away, and lash the rope's ends to the mast itself. If I beg and pray you to set me free, then bind me more tightly still.

The decision to bind oneself is referred to as a "*Ulysses pact*", a pact between two selves. Real-life examples of this abound, from the US Congress trying to find a way to commit itself to reducing State spending, or the decision when opening a new savings account to include predetermined monthly cash-in or to tie it up for a decade. Another example is Dumbledore begging Harry Potter to let him drink a poisonous liquid.

In *Harry Potter and the Half Blood Prince*, Voldemort and his allies are rebuilding their power while Dumbledore is trying to turn the tide, with Harry's

---
[5] Homer, *The Odyssey*, circa 800 B.C.

help. Dumbledore and Harry find the cave where Voldemort has hidden one of the seven parts of his soul. Dumbledore slices his hand with a knife and wipes his blood on a stone to enter the cavern. On an island in the middle of a subterranean lake stands a bowl, at the bottom of which is a necklace containing one seventh of Voldemort's soul. This necklace is protected by a poisonous liquid. Dumbledore knows he has to drink it. He begs Harry to oblige him to drink the liquid to the last drop, regardless of his cries of pain and his demands to stop.

Many authors suggest that human behavior, as illustrated by Dumbledore and Ulysses, results from an internal struggle between "multiple selves", selves that have accumulated as adaptive responses during the process of evolution. Which is the real Dumbledore, the one who cries for mercy, or the one who insists on drinking to the last drop?

Hero But Shy With the Ladies?

Being cautious, taking precautions, being careful with one's health even if the absolute risk of becoming ill seems small, run in direct opposition to the social vocabulary of risk that exists in the world at large. Such vocabulary includes slogans like "*no risk, no reward*", "*just do it*", "*no guts, no glory*", "*no fear*", and "*no pain, no gain*", and encourages us to take real risks with our lives and well-being, even as we continue to flinch every time we encounter an entirely innocuous spider or see the moving image of a snake on the Discovery Channel.

Many characters in movies or in books, and maybe some people around you, display an ambivalent attitude towards risk-taking. John would not take any risk with his savings, took years to approach and court his wife Laura, but does not hesitate to climb a peak 7000 m high in the Himalayas.

In "*Risk taking and personality*", Michael R. Levenson[6], from the School of Public Health, in Boston, Massachusetts, notes that Hollywood scriptwriters portray the Western hero as physically fearless but interpersonally shy, like many of the characters played by John Wayne. This is exemplified when Gail Russell plays the angel (but such a strong woman) and John Wayne the bad man (yet a shy one), in *Angel and the Bad man*, a 1947 Hollywood movie.[7]

Other examples from popular culture include Spiderman, Superman, and Batman—who all struggle desperately with their interpersonal relationships—and certain heroes from children's literature such as Peter Pan and his fairy advisor Tinker Bell, or Pinocchio and his companion Jiminy Cricket. Bram Stoker introduces us to the enigmatic Dracula, whose nighttime and

---

[6] Levenson, R. M., 1990
[7] John Wayne's character in the 1969 movie Rio Bravo was not more confident with the ladies, when confronted with the character played by Angie Dickinson. This double self was an important factor in forging the John Wayne legend.

daytime habits contrasted so interestingly, a character that embodied many of the worries and concerns of the era in which the story was set.

In the same vein, Oscar Wilde wrote *The Picture of Dorian Gray*, the chilling tale of Dorian Gray, the handsome young man who did not wish to grow old, but found a way to keep his looks while everyone around him was beginning to fade. Dorian Gray's other self was a portrait in the attic, a portrait that showed not only how he should have looked by then, but also how he had allowed himself to become morally vile and corrupt. Similar themes are brought out in Robert Louis Stevenson's *Dr Jekyll and Mr Hyde*, which dramatizes the endless conflict between base instinct and culture in the heart of a "civilised" man. All depict multiple selves in conflict.

When Dr Jekyll Becomes Mr Hyde

Heroes from cowboys to Superman struggle to balance contrasting elements of their personalities. Human behavior is full of tensions between our "good" side (Tinker Bell) and our "bad" side, which we try to tame with Ulysses pacts. Dieters pay good money to stay on "fat farms" whose main appeal is that they promise to underfeed their guests; alcoholics take anti-abuse medication which causes nausea and vomiting if they have a drink; smokers buy cigarettes by the pack (rather than by the carton, which is cheaper) because they feel that this may help them to smoke less.[8] We may argue that hiring a personal coach at the gym is a way to acknowledge weakness. But would we do as much exercise without the coach?

Darth Vader, a character from the *Star Wars* movie, displays the typical Machiavellian trait. Machiavellianism, narcissism, and psychopathy, the *Dark Triad Traits*, are described by Peter K. Jonason, Minna Lyons, and Emily Bethell[9] as "*entitlement, superiority, dominance (i.e., narcissism), glib social charm, manipulativeness (i.e., Machiavellianism), callous social attitudes, impulsivity, and interpersonal antagonism (i.e., psychopathy).*" In the *Blank State*[10], Steven Pinker informs us that "psychopaths, who are definitely not "good and kind people", make up about three or four percent of the male population". Psychopaths are however quite extreme folks, not to be confused with our ordinary healthy biased minds: "psychopaths, who lack all traces of a conscience, are the most extreme example, but social psychologists have documented what they call Machiavellian traits in many individuals who fall short of outright psychopathy."

Nevertheless, Star Wars fans discovered that the bad Darth Vader had once been the good Anakin Skywalker. Looking for explanations, some of them

---
[8] Loewenstein, G. H. and Thaler, R. H., 1989
[9] Jonason, P. K., Lyons, M. and Bethell, E., 2014
[10] Pinker, S., 2012

surely concluded that *"it was the loss of his mother as depicted in Star Wars II, Attack of the Clones"* that turned Anakin Skywalker into Darth Vader. In that respect, they would agree with Jonason, Lyons, and Bethell, who claimed that *"Machiavellianism"* could be related to low quality or irregular parental care and relationships. According to the American psychologist Judith Rich Harris, it seems that the influence of parents on their children's personality has been overestimated[11]. Psychoanalysts from Freud onwards have striven to find other kinds of deep-rooted explanations. We shall leave those to them.

The Multi-Modular Mind Hypothesis

The Ulysses pact mentioned above illustrates what Leda Cosmides and John Tooby[12] refer to as *"the multi-modular mind hypothesis"*. Along with other scholars, they asserted that our mind is made up of a bunch of separate modules, or "mental organs", each one adapted to a specific kind of problem, like "avoiding predators", "food searching", "looking for a mate".... So is human behavior the outcome of internal struggles between multiple selves with conflicting preferences?

The philosopher Daniel C. Dennett[13] seeks to explain consciousness with the insights from evolutionary biology, using his *"Multiple Drafts Model"*, which he contrasts with the traditional *"Cartesian Theater"*[14]. According to Dennett, it is hard to get rid of the idea that our brain holds a special center coordinating consciousness, like a unique internal observer. Instead, he proposes that *"at any point in time there are multiple 'drafts' or narrative fragments at various stages of editing in various places in the brain."*

In the register of feelings, the idea of a single emotion system also seems engrained in us. But LeDoux claims that we employ a whole range of emotional devices which have evolved to accomplish specific functions and enable different sorts of feelings. Fear, happiness, shame, and other emotions serve different purposes and provide different solutions to different problems, from avoiding danger to developing fair social relations.

Now, with so many mental modules, we have to choose which things to worry about, because we have only a finite amount of time and brainpower to devote to problem-solving. Life's problems range from finding a spouse to getting a raise from the boss, choosing a tooth brush, or finding our way in a crowd or a forest.

---

[11] Harris, J., 2009
[12] Cosmides, L. and Tooby, J., 2001
[13] Dennett, D. C., 1991
[14] "The Cartesian Theater is a metaphorical picture of how conscious experience must sit in the brain." "According to the Multiple Drafts Model, all varieties of perception—indeed, all varieties of thought or mental activity—are accomplished in the brain by parallel, multitrack processes of interpretation and elaboration of sensory inputs."

Please Alleviate My Cognitive Burden

As Sir Joshua Reynolds noted:

> *There is no expedient to which a man will not resort to avoid the real labour of thinking.*

The Magical Number 7

In folk tales, a hero has to perform three tasks before he can marry the princess, or travel seven seas in order to complete his quest, or the inquisitive maiden learns that she may open six of the doors in her new home, but that the seventh is forbidden.

In real life, as in folk tales, it is often easier when options are limited!

George A. Miller, a psychologist from Harvard University, gave a famous lecture in 1955 demonstrating our cognitive limitations. In his 1956 follow-up paper[15], "*The magical number seven, plus or minus two: some limits on our capacity for processing information*", Miller claimed that our senses and cognitive capacities allow us to distinguish between more or less seven alternatives. As the span of our immediate memory is limited, so is our capacity to memorize and process information. Miller adds that he has been persecuted by an integer, the magical number seven:

> *the seven wonders of the world, the seven seas, the seven deadly sins, the seven daughters of Atlas in the Pleiades, the seven ages of man, the seven levels of hell, the seven primary colors, the seven notes of the musical scale, and the 7 days of the week?*

…not forgetting The Magnificent Seven, the famous 1960 Western movie by John Sturges, featuring the seven actors Yul Brynner, Eli Wallach, Steve McQueen, Charles Bronson, Robert Vaughn, Horst Buchhotz, James Coburn, and Brad Dexter.

When our mind has to grasp anything more than seven items, it tends to package them in easy-to-handle *"chunks"* of information. According to Herbert A. Simon[16], recipient of the Nobel Prize in economics, the psychological reality of the *"chunk"* has been fairly well demonstrated, and the chunk capacity of short-term memory has been shown to lie in the range from five to seven. He states that it takes between 5 to 10 s to record an item of information, a chunk, in the long-term memory. Some other "magical numbers"

---
[15] Miller, G. A., 1955
[16] Simon, H. A., 1982

have been estimated, such as visual scanning speeds and the time required for simple grammatical transformations. Simon believed that short-term memory capacity and the rate at which items can be fixed in the long-term memory are keys to the organization and systematization of both simple tasks and more complex cognitive performances, and explain a wide range of findings.

 An Amusing Test of Short Term Memory

In *The Emotional Brain*, Joseph LeDoux exposes the following experience.[17] Remember this number: 783445. Now close your eyes and repeat it, and then count backward from 99 to 91 by 2 s and try repeating the number again. LeDoux claims that you are unlikely to be able to perform the task.

In fact, once the six figures 7, 8, 3, 4, 4, and 5 are stored in the mental workspace, you have no room left for the operations $99 - 2 = 97$, $97 - 2 = 95$, etc.

So to find more space, and to complete the subtractions, you have to kick the number 783445 out of the working memory. But then once that number has been removed from the mental workspace, you cannot say it out loud again.

The mental workspace in which we temporarily store pieces of information, the so-called "working memory", is limited, and so is the number of items we can hold together, manipulate, and compare in our mind.

"*The memory is full!*" message is not limited to personal computers, smartphones, and digital cameras.

Happiness Is a Matter of (Not Too Much) Choice

Barry Schwartz, Andrew Ward, John Monterosso, Sonja Lyubomirsky, Katherine White, and Darrin R. Lehman[18] suggest in "*Happiness is a matter of choice*" that so-called "*maximizers*" or "*optimizers*" can feel worse as their opportunities increase. One possible explanation is the avoidance of potential regret; the more choices there are, the more likely one is to make a non-optimal choice. A second explanation is that, as the number of choices increases, each seems less attractive, relatively speaking, since there is so much information to deal

---

[17] Ledoux, J., 1998
[18] Schwartz, B. et al., 2002

with. The authors suggest that people may be better off with a limited set of options when they have to choose.

Let's look at a trivial example. You are on a journey in a city. You fancy Italian cuisine and look for a restaurant. Were you a *"satisficer"*, you would pick the first Italian restaurant that pleases you enough in the main street. Now, a *"maximizer"* (optimizer) would try one way or another to gather information to make the "best choice". She could do that by asking around for recommendations, comparing prices and quality, surfing on specialized web pages on the Internet, or buying the *Michelin Guide*.

A Small Dose of Theory On Satisficing

Herbert A. Simon[19] is well known, among other things, for questioning humans' supposed aptitude for behaving as economic optimizers. Simon coined the term *"satisficing"* to refer to when people make a decision on the basis of what is useful enough, and not necessarily what is most useful. As opposed to the optimizers, who tend to look for the most useful choice or the maximal interest.

The Social Number 150?

People are not just individuals. One could even describe the human species as *"hyper social"*. When we interact with others, there is a cost to the brain to live in groups, and to maintain and monitor social relationships on a daily basis. Robin Ian MacDonald Dunbar, a renowned British anthropologist and evolutionary psychologist–head of the Social and Evolutionary Neuroscience Research Group in the Department of Experimental Psychology at the University of Oxford–asked if there was a *"cognitive limit to the number of individuals with whom any one person can maintain stable relationships."*[20]

We owe Dunbar the fruitful discovery that 150, the now famous *Dunbar number*, is more or less an upper bound for the number of social relationships that any given individual can monitor simultaneously.

---

[19] Simon, H. A., 1982
[20] Dunbar, R. I. M., 1993

> **A Small Dose of Theory On the Size of Human Social Groups**
>
> Robin Dunbar used different approaches to get to the figure of 150. Relating the size of the neocortex in primates with their group size, he predicted from the size of the human neocortex *"that humans should live in social groups of approximately 150 individuals."*[21]

Dunbar also looked for typical group sizes in communities, academic disciplines, the army, etc., and observed that figures in the region of 150 to 200 are common in human societies, both old and modern[22]. For instance, in the army, where coordination is essential for survival and success, it is striking to observe such figures for military companies. It seems that, by a process of trial and error, splitting and merging, coordinated human groups have converged to a common range.

Together with Russell Hill of the Department of Anthropology at Durham University, Dunbar examined various social network dimensions in the modern West based on the exchange of Christmas cards. They found that *"Maximum network size averaged 153.5"*, surprisingly close to the 150 deduced from the size of the human neocortex.[23]

So here we stand, with our mind full of *"mental organs"*. Each enjoys its domain of validity, having been tailored for specific tasks. And each displays capacity limitations.

---

[21] Hill, R. A. and Dunbar, R. I. M., 2003
[22] Dunbar, R. I. M., 1993
[23] Hill, R. A. and Dunbar, R. I. M., 2003

The Mind As a Survival Kit

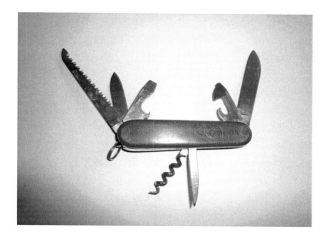

Picture: a multi-purpose Swiss knife

The Mind As an Adaptive Toolbox

Gerd Gigerenzer, a psychologist working at the Max Planck Institute for Human Development in Berlin, Germany, compares the mind to an adaptive toolbox, a bit like a Swiss army knife. Depending on the situation we are in, we can take out and use one or other tool at our disposal, just as the camper with his Swiss army knife can take out a tin opener when he's ready to heat up dinner, or a screwdriver when he realises that there's a screw loose in the camping stove, or a saw to cut firewood, or even a corkscrew to taste an exquisite wine from the French Rhône valley. The most modern form of the Swiss army knife is probably the smartphone, for which the phone function is only one of many goal-specific modules, along with the camera, the flashlight, the maps & GPS, the mp3 reader, a translator, Internet access, sport coaching applications, etc.

According to the Swiss army knife metaphor, the mind has the capacity to adapt depending on the unique circumstances it faces at any given moment.

Savour the astonishing outcome of an experiment by Rodrigo Quian Quiroga, Leila Reddy, Gabriel Kreiman, Christof Koch, and Itzhak Fried, as reported in the scientific journal Nature[24]. They identified *"neurons that are selectively activated by strikingly different pictures of given individuals, landmarks or objects and in some cases even by letter strings with their names"* and, for one

---

[24] Quian Quiroga, R. et al., 2005

of the subjects of the experiments, even … a single neuron triggered by a 1 s snapshot of actress Jennifer Aniston! But it did not fire when the actress was shown alongside her former husband. Other subjects seemed to respond almost solely to different pictures of Bill Clinton.

These types of neurons participate in the identification of people, as well as shapes, and could be made to react to other pictures. The remarkable conclusion of this experiment is that a smaller number of brain cells than was previously thought (similar to a small module) could be involved in face recognition (performed in a fraction of a second). According to Gazzaniga[25], reacting rather like our immune system, our brain holds a palette of neural circuits, and some of them are *"selected out"* and reinforced, when we have to tackle challenges springing from our environment. Of course, we cannot conclude that we were born with a *"Jennifer Aniston recognition neuron"*, but just that some individuals have specialized their neural circuits by intense training, for example, passionately watching *Friends* on TV, or browsing magazines in the dentist's waiting room.

Our take here will be to view the brain as a kind of survival kit filled with inherited *"problem solving devices"*, such as being able to distinguish a snake from a stick, discerning a friendly neighbor from an enemy, a poisonous mushroom from a savory *Boletus edulis* (also known as "*cèpe de Bordeaux*").

Our Biases Reflect Human Ecological Rationality

As we will see, many of our mental biases are responses shaped by natural evolution and adaptation. Some biases are functional distortions of reality—it's better to take a stick for a snake than the contrary—with the consequence that truth is not necessarily aligned with fitness, the currency of evolution.

> *There are many situations (…) in which it can be adaptive to distort reality. Even massively fictitious beliefs can be adaptive as long as they motivate behaviours that are adaptive in the real world*[26]
> David Sloan Wilson, Professor of Biological Sciences and Anthropology at Binghamton University

Hence, some of our biases still make sense in the modern world. Like our caveman ancestors, we still strive to have kids and, sometimes, to avoid danger and survive. Biases can be useful here, leading us to contemplate objects from a new angle that provides an advantage in terms of survival and reproduction.

---

[25] Gazzaniga, M. S., 1992
[26] Wilson, D. S., *Darwin's Cathedral,* The University of Chicago Press, 2002

For instance, as we will see later, overestimating the height of a cliff from the top (with respect to a view from the bottom) reduces the risk of falling, and assessing an approaching vehicle as closer than it really is, may save your life.

Some biases can also be responses unfit for our modern world. Indeed, our social and physical environments have changed considerably and evolution has not had time to reshape our bodies and minds, so we are sometimes left with behavioral tendencies that no longer always make rational sense. The taste for fat—as in mouth-watering hamburgers, French fries, and ice-creams—enters into that "unfit" category. In our wealthy modern society, where there is no shortage of food, this can induce health problems. It is an example of a maladaptive response of our organism. Our leaning toward fatty foods (such as the savory French cheeses below) was an adaptive bias in the old days, when food supply was uncertain. Indeed, when food was available for our hunter-gatherer ancestor, it was an advantage to store it as body fat, as a buffer for more difficult times.

Picture: three French cheeses *Cantal, Saint Nectaire,* and *Fourme d'Ambert*

Today, some of our instinctive responses may not always be appropriate to our modern needs, and indeed they may sometimes be in direct conflict with them. The comfortable life we live in the twenty-first century has been brought to us largely by technology. Yet, as our world has moved on, our instincts have not kept pace with the many changes in our environment, which means that we have no choice but to react with our sometimes out-of-date mental "hardware". As claimed in *The Mystery Method*[27], we are *"prisoners in time"* with a brain hard-wired for what psychologist John Bowlby called the *"environment of evolutionary adaptedness"* (in his 1969 book *"Attachment and Loss"*), and a set of *"softwares"* that is not always up to date.

> *We cannot rely on intuitions honed by our everyday experiences in the modern world ... behavior generated by mechanisms that are adaptations to an ancient way of life will not necessarily be adaptive in the modern world.*
> Jérôme Barkow, Leda Cosmides, and John Tooby[28]

The viewpoint of "ecological rationality" is advocated by various scholars, including John Tooby, Leda Cosmides, Gerd Gigerenzer, and others. Such a perspective can help us to understand our modern consumer patterns. In the *Journal of Consumer Psychology*, Vladas Griskevicius and Douglas T. Kenrick, from the University of Minnesota and Arizona State University[29], respectively, analyze the underlying motives for consumption and choice from an evolutionary perspective. Based on a review of evidence, they list *"(1) evading physical harm, (2) avoiding disease, (3) making friends, (4) attaining status, (5) acquiring a mate, (6) keeping a mate, and (7) caring for family."* Consciously or not, our aspirations reveal deep and evolutionarily meaningful ends.

By understanding how our minds—and those of others—work, we can combine instinct and reason to make decisions and act more effectively than either approach can achieve on its own. Read on to find out why the human mind, which developed over the course of many millennia of hunting-gathering and cave-dwelling, is sometimes poorly adapted to cope with the maelstrom of choices, influences, and experiences that assail us in our daily lives.

---

[27] Mistery, 2007
[28] Barkow, J., Cosmides, L. and Tooby, J., 1992
[29] Griskevicius, V. and Kenrick, D. T., 2013

3
Better Be Paranoid to Survive

Ethologist and evolutionary biologist Richard Dawkins has coined the term *"life-dinner principle"*[1] after Aesop's fable according to which

the rabbit runs faster than the fox,
because the rabbit is running for his life,
while the fox is only running for his dinner.

Here we enter the territory of our hunter-gatherer ancestors in the savannah. It must have been rather scary, out there in the open, exposed to the elements, in the presence of creeping spiders, crawling scorpions, and slithering snakes, not to mention a host of other wild creatures inspiring fear and generating risk. Like the rabbit in the life-dinner principle, our ancestors had to fight to survive. Maybe they were better off being a little paranoid. So meet the paranoid optimist!

Pictures: between the devil (a hungry lion in the savannah) and the deep blue sea (poisonous jellyfish)

---
[1] Dawkins, R. and Krebs, J. R., 1979

It Was Scary In Flintstone…

Why are we more scared of spiders than cars or electric sockets, even though most of us see few spiders on a daily basis?

Picture: Aztec skull sculptures in a temple (Mexico City)

Picture: a fine collection of dead spiders, scorpions, and centipedes

The World Is Populated With Survivor's Heirs

It is a simple fact that we are surrounded by organisms that are the heirs of survivors. In days gone by, only those who could escape lethal risks survived to tell the tale. Not to put too fine a point on it, when confronted by an angry lion or cornered by a furious woolly mammoth, the person who reacted quickest and most appropriately was the one who got to pass on his or her genes to the next generation.

In one form or another, we undoubtedly possess some of the abilities that allowed our predecessors not to dither in the face of danger, while the others were simply eliminated.

The world is full of living beings, including ourselves, whose ancestors succeeded in surviving and reproducing. With each generation, the ones who managed to survive the dangers in their environment were sometimes able to procreate, while the others did not, or only did so less successfully or less frequently; that's evolution. In the process, the winners' brains were selected, their minds adapted to recognize and escape from dangerous situations.

Today, the biases in our minds reflect the strategies that kept our successful ancestors alive, because their genes have been transmitted to us, their descendants. In other words, we have evolved to be good at getting out of difficult situations and producing progeny who will carry our genes into the future.

Our Hunter-Gatherer Parents Thrived In the Savannah

Picture: lake Awasa in the Ethiopian savannah

Research suggests that the population of Homo sapiens began to expand dramatically during the Late Stone Age, about 40,000 years ago,[2] although temporary settlements have been reported in caves in the Dordogne in France dating to about 100,000 years ago. From the Wurm period until the

---
[2] Leakey, R. E. and Lewin, R., 1977

beginning of the Neolithic about 9000 BC, our species must have been exposed to a fairly stable set of conditions—homogeneous fauna and flora over very long periods of time, low human population density, long periods of stable climate (either warm spells or ice ages), plagues and epidemics, and high infant mortality.

Before that, humans survived for countless generations as nomadic hunter-gatherers. During that early period, the risks they faced most commonly were a lack of food, wounds and death incurred while hunting, avoiding predators who might want to hunt and eat *them*, and disease. David Buss[3] lists as potential threats what Darwin called "*hostile forces of nature*": climate, weather, food shortages, toxins, diseases, parasites, predators, and hostile conspecifics.

Early settlements show the remains of stones, bones, horn, or wood, which were the primary materials used in making tools, dwellings, and weapons. Throwers, arrows, and harpoons were used to hunt large mammals. The importance of these large animals is attested by the beautiful wall paintings that survive in caves to this day, and which typically depict the concerns of the people of those times, but not necessarily the animals that they really hunted (buffalos, lions, bears, and horses, etc.). In this, they resembled all other living creatures: as Charles Darwin said in *On the Origin of the Species*, "*Fear of any particular enemy is certainly an instinctive quality, as may be seen in nestling birds*".

Picture: cave paintings at Pech Merle (Cabrerets, Lot, France), partial view of the dappled horse panel, ©P. Cabrol and Center of Prehistory of Pech Merle

---

[3] Buss, D., 2014

The Neolithic saw a fundamental change in how our ancestors lived, and while the cultural changes were dramatic, they occurred over a relatively short period, not all that long ago. This was when the practice of agriculture began, soon becoming the norm throughout many human groups. As a result of agriculture, people started to settle down, remaining for long periods in the same place. Now they could create a surplus for lean times, sowing the seeds, both literally and figuratively, for the development of economics and banking as they exist today. With the development of agricultural stocks, war and other threats began to manifest themselves more frequently. And agriculture and settlement created associated risks: denser populations led to the easier spread of disease and pests such as disease-bearing fleas, rats, and mice. In good years, agriculture could lead to a surplus, but in bad years, famine stalked the land and people died in their droves. Slightly later on, when people started to use metal, this also brought in new risks.

In summary, what we know as "history" is but a fraction of our past as human beings, as Homo sapiens. Our prehistory stretches back much, much further and, in evolutionary terms, it has surely made a much bigger difference to the way we are today. Our behavioral patterns and reactions to potential risk have been inherited or learned from our species' hunter-gatherer past. As mentioned before, such risks include dangers from wild animals like snakes and spiders, as well as natural elements like rain and cold which create the need for shelter, and from man-made factors such as weapons. As a result, the things that we instinctively *feel* to be riskiest (although the rational mind may understand that they actually only pose a low level of risk today), the things that we are most scared of, are usually the ones that scared our hunter-gatherer ancestors the most, while we tend to be less concerned about the many real risks associated with our modern life. The latter might include atmospheric pollution, newly discovered viruses, modern weapons, and food contamination, for example. After all, the impact on human evolution of 10,000 years of farming and a short industrial period is a lot smaller than the unimaginable number of years we spent on the move, relying on the animals we could kill and edible plants we could find.

According to Joseph Ledoux, *"fear"* is the system that detects danger *"and the behavioral, physiological, and conscious manifestations are the surface responses it orchestrates"*.[4]

Will the understanding of how and why our fears act upon us, help us to make more rational decisions in the course of our daily lives? We do not know for sure, but it must surely be worth acquiring such knowledge!

---

[4] Ledoux, J., 1998

SSSSSS...Sinuous Snakes Still Scare!

... wretched by the death of thee, Than I can wish to adders, spiders, toads, Or any creeping venom'd thing that lives!
—William Shakespeare, King Richard III, 1592

Great holes secretly are digged where earth's pores ought to suffice, and things have learnt to walk that ought to crawl.
—Howard P. Lovecraft, The Festival, 1925

A famous quote from the French playwright Jean Racine in his masterpiece *Andromaque*, repeats the «S»:

Pour qui sont ces serpents qui sifflent sur vos têtes
(For whom are these snakes that whistle on your heads).

Picture: a green python

Today, when people spontaneously recoil from the very idea of snakes and spiders—despite the much greater risk posed today by passing cars or electrical appliances—they reflect our hunter-gatherer heritage. We are all responding to the risks that humans evolved to recognize over thousands and thousands of years.

Snakes hold symbolic importance in many religious beliefs and rites all over the world, although most of us see snakes rarely, and then safely emprisoned in a vivarium at the zoo. Christian tradition often represents the devil as a snake, while the snake of an ancient Aztec myth is featured on the national flag of Mexico. To this day, groups who wish to represent themselves as outsiders or rebels, often integrate the snake into their iconography. Just look at the millions of heavy metal posters or Hell's Angel jackets that use snake imagery.

John Tooby and Leda Cosmides point out why so many of us still fear snakes today in their fascinating 1996 paper "*Are humans good intuitive statisticians after all.*"[5] Over evolutionary time, encounters with snakes resulted statistically in a certain proportion of lethal bites.

Picture: a snake sculpture at the Anthropology Museum of Mexico City

Picture: snakes from the Forbidden City in Beijing

Those humans who carried an effective "snake avoidance module" stood more chance of making it to the reproductive stage than those who did not. And we are of course their great … great … grandchildren. This is why today a high proportion of humans still remain afraid of snakes, even if few of us have ever encountered one that could have bitten us.

---
[5] Cosmides, L. and Tooby, J., 1996

Picture: a Chinese dragon from the Summer Palace in Beijing

Cosmides and Tooby extended their observations to a discussion of how wild animals are perceived. They report the study of psychologist Adah Maurer who observed that *"modern, urban children generally fear things that actually pose very little risk to them"*, namely wild animals.[6] Modern children rarely wake suddenly at night after a bad dream about obesity, although this is a very real threat to many young people's health in the modern age. We are yet to meet a child with a fear of potato chips and fizzy drinks, although over-consumption of just these products is currently damaging the present and future health of untold millions of children around the world. On the contrary, they are much more likely to have bad dreams about sharp-toothed animals such as sharks or crocodiles, which most children today will only ever encounter in the zoo or on the Discovery Channel:

> *In fact, most of the five- and six-year-olds in Maurer's study of Chicago schoolchildren mentioned wild animals (most frequently snakes, lions, and tigers) in response to the question, "What are the things to be afraid of?" Only older children, from the age of twelve, gave individual replies that related to wars, bombs, or social concerns.*[7]

---

[6] Cosmides, L. and Tooby, J., 1996
[7] Cosmides, L. and Tooby, J., 1996

...It's Scary Now!

We will find again many of these ancient fears in modern life and explore a few dimensions of risk in recent history.

What Makes a Landscape Friendly?

Children not only fear absent lions, they also prefer pictures of landscapes that they have never experienced, especially savannah-type landscapes, as demonstrated by John D. Balling and John H. Falk's experiments:[8]

> *Overall, savanna and open forest scenes tended to be highly preferred, while the thick forest or jungle and desert slides were clearly disliked. (...) The strongest preference for savanna was found among the two youngest age groups.*

In surveying a particular landscape and pronouncing it "beautiful", the viewer is looking at the shapes and forms laid out before him and, whether accurately or not, interpreting them as representing environmental conditions that were once favourable to survival. This might include the ability to see for miles, or the prospect, in combination with the ability to hide, represented by a refuge. Such findings are show-cased in *Evolved Responses to Landscapes*,[9] by Gordon H. Orians and Judith H. Heerwagen and in *Environmental Preferences in a Knowledge-Seeking, Knowledge-Using Organism*, by Stephen Kaplan, who reports rating reaction times of 10 ms. When survival is at stake, it's the first impressions and the quickest that matter!

 A Small Dose of Jay Appleton's Landscape Theory

In 1975, British geographer Jay Appleton published a book called *The Experience of Landscape*,[10] which developed his theory that human aesthetics are predicated around our need for both opportunity and refuge, in particular by assessing the presence or absence of risk.

The prospect of a wide open horizon or the mere existence of a point of view allowed us to make sure of predators' absence, as well as to spot the presence of water, or of food (wild animals, fruit trees).

A cave, a cavity, or a cliff, were natural refuges where no predator, nor enemy, could attack us from behind.

---

[8] Balling, J. D. and Falk, J. H., 1982
[9] Barkow, J. H., Cosmides, L. and Tooby, J., 1992
[10] Appleton, J., 1975

We see echoes of Appleton's theory in contemporary advertising for golf courses, which are typically located in areas considered to have great scenic beauty, and which are great examples of landscapes that combine opportunity with safety—prospect and refuge.

Picture: undisturbed sky reflections in a pond

A recent study by Patrick Hartmann and Vanessa Apaolaza-Ibáñez[11] reveals adult *"preferences for images of lush green landscapes with water and familiar biomes"*. Here we have a potential application of evolutionary psychology to advertising!

Picture: a soothing landscape in the South West of France

Why does contemplating the sea from the shore seem relaxing to so many people? Is it the presence of water, a welcome resource in the savannah? A wide-open perspective, free of predators?

---

[11] Hartmann, P. and Apaolaza-Ibáñez, V., 2010

Between the Rock and a Hard Place

Rocks, currents, no anchorage, sheer cliff to lay to, no insurance company would take the risk …
Lord Jim, Joseph Conrad.

Reefs stir up feelings of risk. The authors carried out a survey of several museum collections that revealed that few if any pictures carry titles that incorporate "risk", although a few mentioned "fright" or "fear". Nonetheless, some noteworthy works of art display the risks faced by humans.

Fishermen at Sea, by the famous painter Joseph Mallord William Turner (1775–1851), evokes the threat of rolling waves, storm clouds, and the treacherous "Needles" rocks in the background, just off the Isle of Wight.[12]

It is said[13] that the word "risk" itself comes from the ancient Italian "*risicare*", meaning "*to dare*". Others say that the word risk originates from the term "*reef*", things that cut, from the days when reefs could cut through a ship at sea.

Another of Turner's paintings, *The Wreck of the Minotaur*, illustrates that reef etymology of the word "risk".[14]

Picture: shipwreck of the Minotaur, circa 1810, by J. M. W. Turner

---

[12] http://www.tate.org.uk/art/artworks/turner-fishermen-at-sea-t01585
[13] Bernstein, P. L., 1998
[14] http://www.minotaur.org/minotaur-turner.htm

According to Wikipedia, *"whilst sailing from Gothenburg to Britain, under the command of John Barrett, the Minotaur struck the Haak Bank on the Texel off the Netherlands in the evening of 22 December 1810".*

That dramatic event and the trial that followed were largely publicized at the time.

> *The customary court martial decided that the deceased pilots were to blame for steering the ship into an unsafe position, having misjudged their location by over 60 miles because of the weather.*[15]

In the modern world, an entire industry—the business of insurance—has been built up around risk and efforts to minimize the costs it incurs. While it was initially developed to hedge the uncertainties of sea journeys involving the transport of merchandise, one can now purchase insurance for most risky activities, from bungee jumping to parenthood, and even on behalf of our pets, should they fall ill and require expensive veterinary treatment.

Insurance shares the risk among those who are spared. And this explains where the word "average" comes from. In navigation, one sometimes had to throw excess weight overboard to save the rest of the cargo; losses due to "*avaria*" were compensated on the basis of the value that was preserved, with a factor known as "*average*".

Fear emotions associated with "things that cut", like sharp teeth or nails, have been exploited by movie directors from Steven Spielberg (Jaws) to Jackie Earle Haley (Wes Craven's character Freddy).

Picture: a confident man riding a motorbike at full speed while carrying a pane of glass

---

[15] http://en.wikipedia.org/wiki/HMS_Minotaur_%281793%29

In modern Chinese characters, the word "*risk*" is written with two symbols. The first means "*wind*" or "*breath*" and the second means "*danger*". Risk, therefore, can be considered as the breath of danger, or, in a second interpretation, "*wind*" is itself ever-changing and uncertain, so that the word risk means both uncertainty and danger.

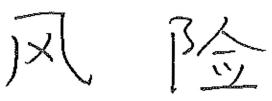

Picture: risk in modern Chinese[16]

However, both symbols above derive from traditional writing, in which the word risk is written with two symbols. The first symbol comprises two animals protected from the wind by a veil. The second represents two men, protecting themselves and their food by a roof and an earth wall.

Proverbs Convey Cultural Risks

Elke U. Weber, Professor of Psychology at Columbia University, Christopher K. Hsee, Professor of Behavioral Science and Marketing at the University of Chicago, and Joanna Sokolowska, from Academy of Science, Warsaw, Poland, demonstrated that the majority of respondents in four cultures (Germany, United States, Poland, and China) were perceived to be risk-averse, showing a strong disinclination to take risks.[17] We can safely assume that a tendency towards risk aversion is a general feature across humanity as a whole, regardless of the underlying differences in culture, and despite the fact that we also have a tendency to enjoy playing with risk, and even admire prominent risk-takers.

---

[16] Special thanks to 收件人: Yuexi Ma
[17] Weber, E. U., Hsee, C. K. and Sokolowska, J., 1998

However, that is not to say that the picture is the same, regardless of where one looks. Risk perception may differ from one culture to another according to the protection ("*cushion*" factor) against possible negative outcomes, which may vary. People who live in countries with social welfare programmes, for example, may behave somewhat differently from those who do not, in the face of certain risks.

Weber and Hsee's results suggest that, regardless of the nationality of those asked to rate proverbs, Chinese proverbs seem to encourage greater risk-taking than American proverbs. The latter include:

> "A bold attempt is half success"; "Aim at the stars but keep your feet on the ground" and "He that fights and runs away may live to fight another day."

Compare with the Chinese proverb:

> "If someone has never left his home, he cannot be a great person."

Picture: a gigantic bronze spider in the hall of the Doha exhibition center during the Climate change conference in November 2012

Paris In the Summer When It Sizzles[18]

You might say that risk is what sends shivers down your spine and makes your blood run cold, but how easy is it really to identify what is risky and what isn't?

In 2010, the authors were involved in an AXA-Paris School of Economics research project.[19] The study was based on two focus groups[20] of twelve people, each group including six carefully selected investors and six other participants who were non-expert investors. They were exposed to images that might trigger different possible feelings of risk. The participants were asked to select three images among nine that really inspired such a feeling of risk to them. The "riskiest" items, chosen in preference to the others, were the revolver, the tightrope walker, the syringe (with its needle), and the wrecked ship. Symbolic visuals such as the red triangle (suggesting alarm and instability, as we will discuss below), bank logotypes (suggesting bank institutions and power), Napoleon (suggesting war—this was a *French* group after all!), and banknotes (suggesting money) were apparently less vivid than concrete dangers. The items that were chosen were "easy to understand dangers" that could harm anyone on a personal level (drowning, pain, or falling). In each case, the images that were chosen implied a risk of death.

Needles Can Breach the Body's Defences

So, needles are scary. No surprise. Our ancestors feared the sharp claws of wild beasts tearing their skin.

---

[18] Cole Porter song "*I love Paris*". Lyrics extract:
  "I love Paris in the springtime.
  I love Paris in the fall.
  I love Paris in the winter when it drizzles,
  I love Paris in the summer when it sizzles"
[19] Research project AXA—PSE entitled "The economics and psychology of risk taking, impatience and financial decisions: confronting survey, experimental and insurance data".
[20] A focus group is a widely used qualitative research format in which a selected group of people are asked about and may expand upon their perceptions, opinions, beliefs, and attitudes. These focus groups were conducted in Paris between September and December 2010. The questions were originally presented in French, and have been translated for the purpose of this book.

Picture: a nightmarish syringe in the sand

Aside from *trypanophobia,* or fear of medical procedures involving needles or injection, there are those who suffer from a fear of sharp and pointy things in general, even pointing fingers. A TV series, *Monk*, focuses on the life and times of the character Adrian Monk, a one-time detective and now consultant for the San Francisco Police Department who suffers from a number of psychological disorders, including obsessive-compulsive disorder and several phobias. In season five, two of the characters attend a rock concert and discuss an individual called Stork.

> *"Well, Kendra, we were there," Natalie says. "We saw a needle in his arm."*
> *"That's how I know something's wrong," her friend responds.*
> *"Stork was completely phobic about needles. He was the only roadie I ever met that didn't have a tattoo.*
> *I mean he missed a whole South American tour last year because he wouldn't get vaccinated!"*

On being asked what he thinks, Adrian Monk rejoins that one doesn't just get over a phobia. Uncountable numbers of needle-phobic people today would concur.

Of course, injection also stands as a potent sexual symbol. And Paris HEC Marketing Professor Jean-Noël Kapferer,[21] who doubles as a specialist in rumour-mongering and urban legends, tells of "*injection epidemic*" rumours that were common in Paris in 1820. And again, in December 1922, there were completely unfounded reports that the customers of major Paris stores were being "pricked" by mysterious madmen as they went about their shopping.

---
[21] Kapferer, J. -N., 1990

We Are Scared of Transgression

In a bold shortcut, let us proceed from breaches of the skin to modern transgressive risks such as nuclear energy, shale gas, and Genetically Modified Organisms (GMO).

Combining the fear of sharp objects, the fear of wild animals, and wild sexual fantasy, vampire stories from Bram Stoker's *Dracula* to its many modern spin-offs, including the contemporary bosom-heaving *Twilight* series, tell tales of being bitten by vampires. These stories have been frightening and titillating a large audience for generations. The bite pierces the (often) virginal heroine's skin, damaging her and at the same time, releasing her from the constraints of society, to become one with danger.

Urban legends are a rich source of stories that illustrate innate concepts of risk and danger. You may have heard the one about the lady who went on holiday to an exotic location. On return, she found that she had developed an unpleasant rash that resisted all treatment. Eventually, a boil was lanced and … a host of newly hatched insects emerged from under her skin!

Modern urban legends on this theme tell how children are condemned by the bite of dangerous insects hiding in an apparently innocuous object like a flower pot or toy. Breaching claws still haunt our darker moments.

Surprisingly, we found the same idea of transgression in areas where one would not have naturally thought. Nuclear energy and GMO technologies stand as "*modern transgressive claws*".

Could it be that reactions to GMOs and nuclear energy have something to do with man transgressing Nature by entering deep into its essential building blocks, the cells and the atom, as if irresponsible scientists had somehow dared to cross a "natural skin"?

Peter Sandman,[22] a risk communication speaker and consultant, observed that if you make a list of environmental risks in order of how many people they kill each year and then list them again in order of how alarming they are to the general public, the two lists will be very different. The risks that kill are not necessarily the risks that anger and frighten. Sandman suggested adding a moral dimension to physical hazards. Outrage and other moral considerations do not constitute a bias in the way we perceive risk, but represent part of our very conception of it.

According to Sandman's formula, "*Risk = Hazard + Outrage*", adding a moral dimension to the concept of physical hazard can improve our understanding of risk perception.

This is supported by Lennart Sjöberg,[23] Professor Emeritus in the Department of Marketing and Strategy at the Stockholm School of Economics, who

---

[22] Sandman, P. M., Miller, P. M., Johnson, B. B. and Weinstein, N. D., 1993
[23] Sjöberg, L., 2000

points out that a substantial dimension of risk perception has to do with features like "unnatural" and "immoral". In the psychological paradigm, Paul Slovic, professor of psychology at the University of Oregon, and his colleagues propose the following risk perception factors as affecting attitude:

1. Dread
2. Uncontrollable
3. Unnatural
4. Unknown
5. Feared
6. Catastrophic
7. Unjust
8. Confidence

Among other risk dimensions, the "*dread*" psychological factor relates to a perceived lack of control, and the "*unknown*" psychological factor, the extent to which hazards are judged to be unobservable, unfamiliar, and delayed in impacts.[24]

Sjöberg claims that the explanatory power of all the other factors almost doubles when "unnatural risk" is included. Hence, to assess factors relevant in risk perception, it is important to look for features revealing some kind of interference with nature—these may include tampering with nature in some way, a feeling of immoral and unnatural risk, or the display of human arrogance.

People are more upset by diseases that they believe result from "*immoral*" behavior or "*unnatural*" actions. During the outbreak of Mad Cow Disease (bovine spongiform encephalopathy), originating in the United Kingdom in the 1980s, the fact that naturally vegetarian cows were fed with meat by-products, causing them to develop the disease, contributed much more to the public's symbol-laden assessment of the situation than the actual mortality rates.

We Have Not Been Hard-Wired to Cope With Modern Risks

We are surrounded by technology designed to make our lives easier than ever: smart phones, GPS, induction hobs... We are living longer, healthier existences than our ancestors could ever have imagined possible. In developed countries, many of us benefit from lives rich in opportunities and choices.

Since the Industrial Revolution, humanity has been confronted with wholly new risks that spring from human intelligence and creativity, for better and for worse: pollution, world wars, nuclear proliferation, terrorism, global change, etc. These man-made risks, brought about by technology and science,

---
[24] Fischhoff, B., Slovic, P., Lichtenstein, S., Read, S. and Combs, B., 1978

are more diffuse and generalised than those confronted by our hunter-gatherer ancestors, and our caveman brains have not been wired to react to them, compelling us to default to our standard psychological settings.

Therefore, today's risks tend to be vaguer, more difficult to pin down than, let's say, the threat posed by a sabre-toothed tiger on the prowl for some fresh meat for dinner. The appearance of the modern threat, its scope, and the speed of the changes it may involve are all unknowns.

Unknown New Risks Surround Us

The modern world has helped and still helps humanity to avoid risks such as diseases, epidemics, and starvation, among others. It also exposes us to a broad range of risks in various domains such as health, environment, transport, industry, technology, finance, etc. According to German sociologist Ulrich Beck,[25] while people learned how to protect themselves, as individuals, from external threats—by making weapons to protect them, building housing to shield them from the rain and cold, and slowly accumulating the knowledge needed to interpret their environment, which was consistently hostile— they feel unable to confront exponential and unknown industrial, scientific, and technological innovations which threaten humanity as a whole.

Technological development brings new, emerging, controversial risks that the modern caveman has not yet evolved sufficiently to grasp, at least not with the same ease, from cellular phone radiation to genetically modified organisms, contamination, and the impacts of climate change; risks that are unknown or still unknowable in terms of their scope and reach.

Unknown Unknowns Are Difficult to Pinpoint

And beyond unknown risks lies what former United States Secretary of Defence Donald Rumsfeld evokes in a speech made on February 12, 2002, discussing the vertiginous issue of having to take decisions in the face of uncertainty:

> *Reports that say that something hasn't happened are always interesting to me, because as we know, there are known knowns; there are things we know we know. We also know there are known unknowns; that is to say we know there are some things we do not know. But there are also unknown unknowns, the ones we don't know we don't know. And if one looks throughout the history of our country and other free countries, it is the latter category that tends to be the difficult ones.*[26]

---

[25] About new risks, see: Beck, U., 1986; Dupuy, J. P., 2000
[26] http://en.wikipedia.org/wiki/There_are_known_knowns

The expansion of mankind on the planet also brings its share of risks, ranging from the exposure to flooding that results from the construction of housing in vulnerable regions to proximity with reservoirs of unfamiliar viruses when we penetrate areas that have never previously been inhabited by humans. The unknown unknowns raise delicate issues when we have to make decisions for the future.

After such a catalogue of risks, who would not turn paranoid?

Are We Paranoid Optimists?

Well listen up. There is a storm coming like nothing you have ever seen. And not one of you is prepared for it.
Do you think I am crazy?

From the movie *Take Shelter*, directed by Jeff Nichols, 2011

Pictures: a summer storm in Pinamar Argentina, 2015

Movie director Alfred Hitchcock was a world expert in striking the chord of fear. Inspired by a Francis Iles novel, the Alfred Hitchcock movie *Suspicion* tells the story of a woman who suspects her husband of being a murderer. By night, Johnny (played by Cary Grant) walks up the stairs with a glass of milk (lit from within)[27] that he offers to his wife. Though suspicious of being poisoned, Lina (played by Joan Fontaine) takes the risk and drinks. In other words, Joan Fontaine's character took the risk of drinking despite her doubts. And so, let us take the step from suspicion to paranoia.

Sometimes It Is Best to Be a Little Paranoid! (Paranoia Saved Our Optimistic Parents)

The biased mind is tuned in favor of a useful paranoia, as smoke detectors are tuned towards false alarms—better to be disturbed by false alarms than to be trapped in a blazing inferno, right? This is the mind-blowing perspective that Martie G. Haselton and Daniel Nettle developed in their "*Paranoid Optimist*" theory. Well, we have slogged through their research and extracted a few gems for you.

---

[27] In a series of interviews with Alfred Hitchcock by François Truffaut (Gallimard, 1993), Hitchcock tells that he had put a light in the glass of milk to make the viewer look at the glass

Louder Is Closer

Our perceptions display asymmetry. Sounds that become progressively louder seem to be closer than they really are. For example, people generally overestimate the proximity of approaching vehicles. This fascinating observation has been made by Professor John G. Neuhoff, from the Department of Psychology of the College of Wooster in Ohio.[29] Neuhoff has conducted psychoacoustic studies, brain imaging studies, sex difference studies, and comparative studies with Rhesus monkeys. His research has demonstrated that, when people hear approaching sounds, they tend to underestimate how long it will take the person or thing making the sound to reach them.

Why do increasing sounds seem to creep so much closer?

Once again, we can look to the circumstances in which our ancestors' brains evolved. Assuming that the origin of sounds is closer than it actually is would be a useful bias when a predator is approaching. The person who believes that

 A Small Dose of the Paranoid Optimist Theory

In their 2006 paper, *The Paranoid Optimist*,[28] Martie G. Haselton, Professor at the Departments of Communication Studies and Psychology, University of California, Los Angeles, and Daniel Nettle, Professor of Behavioural Science at Newcastle University, provided a startling picture of the human mind.

Haselton and Nettle claim that natural selection has favoured people with a bias towards false alarm in situations displaying asymmetry between costs.

a hungry tiger will arrive in 30 s will flee more quickly, and is more likely to be in a safe place when he ambles along a full 2 min later! In agreement with Neuhoff's observations, this bias is likely to be adaptive, because it is better to be prepared too early for something dangerous than too late (the degree of preparedness depending on the relative costs of protection and danger). Natural selection favours neural mechanisms that detect approaching sounds in a way that makes them appear closer than they are, in a manner that is asymmetric with respect to receding sounds. This is the familiar principle of

---

[28] Haselton, M. G. and Nettle, D., 2006
[29] Neuhoff, J. G., 1998

the smoke detector: it is better to tune a smoke detector so that it will always detect a genuine fire, even if the cost is the occasional false alarm.

Our perceptions do not only display sound asymmetries. Research has also shown that approaching objects seem larger and brighter[30] and that heights are not perceived equally from above or below.[31]

Descent Illusion

You go to Paris and visit the Eiffel Tower. You look at it from the ground and are impressed by its height and majesty. Then, you take the elevator and go to the top. Now, from the top, you look down to where you were before, way below (in the old days, you were allowed to look from the fence, but from September 2014 you can walk on a glass pane on the first floor, 115 m above the ground). Wow! That looks a long way down! The distance looks much greater from the top than from the bottom: this phenomenon has been coined the *"descent illusion"* by Russell E. Jackson and Lawrence K. Cormack, from the University of Texas, Austin.[32] They observed that participants in experiments perceived greater vertical distance when viewing from the top than when viewing from the bottom, the increase ranging between 32 and 84 %. Jackson and Cormack suggest that *"natural selection has differentiated some psychological processes, including height perception, in response to the navigational outcome of falling"*. Because *"descent results in falls more often than does ascent"*, evolution has shaped our perceptions so that, by an adaptive illusion of height increase, we behave more carefully at the top of a cliff, a hill, or a tree.

Remember James Stewart's character Scottie in Hitchcock's Vertigo. Here is an extract from the 1957 script by Alec Coppel and Samuel Taylor, showing the gut feelings associated with vertigo.

> EXT. SAN FRANCISCO ROOF TOPS—(DUSK)—MED. SHOT
> We now see a short gap between rooftops, with a drop below.
> The pursued man makes the leap successfully followed by the uniformed policeman. Scottie makes the same leap, but almost trips in taking off and is thrown off balance. He tries to recover, lands awkwardly on the opposite roof, and falls forward, prone, with a heavy impact that hurts and drives the breath from his body. He tries to rise but raises his head with a look of pain—one leg is doubled up under the other. The tiles give way, and he slides backwards, and his legs go over the edge of the roof, then his body. In his daze he grasps at

---

[30] Sutherland, C. A. M., Thut, G. and Romei, V., 2014
[31] Jackson, R. E. and Cormack, L. K., 2007
[32] Jackson, R. E. and Cormack, L. K., 2007

the loose tiles, and as he goes over the edge he clutches on to the gutter, which gives way, and he swings off into space, looking down.
EXT. SAN FRANCISCO ROOF TOPS—(DUSK)—CLOSE SHOT
Scottie looking down.
EXT. SAN FRANCISCO ROOF TOPS—(DUSK)—LONG SHOT
From Scottie's viewpoint, the gap beneath the building and the ground below. It seems to treble its depth.
EXT. SAN FRANCISCO ROOF TOPS—(DUSK)—CLOSEUP
Scottie looking down with horror. His eyes close as a wave of nausea overcomes him …

Picture: rock formation known as "*La Roche de Solutré*", in Burgundy, France (also an important prehistoric site) We did not dare to climb the hill and look at the cliff from the top, for fear of vertigo!

Our Radar Minds Are Tuned Like a Smoke Detector

Look at the familiar principle of the smoke detector: it is better to tune a smoke detector in such a way that it always detects a genuine fire, even if the cost is the occasional false alarm. We put up with our smoke detectors occasionally going off when we are just making toast or "*crème brûlée*" because we know that that's better than having a sluggish alarm that doesn't sound when there is a real fire that poses a serious threat to life.

3 Better Be Paranoid to Survive   **43**

Picture: a snake-like wood branch

Imagine one of your ancestors walking through the grass when suddenly he sees it moving ahead. He stops. Is it a snake? He doesn't know, so he's not sure whether to go straight ahead or make a detour to avoid the possible threat. Even if it *is* a snake, is it dangerous or not? The moving grass doesn't give enough information to tell. If it is just the wind, or maybe a harmless dormouse, the detour is a waste of time. But if there *is* a snake and he goes ahead, the risk of being bitten is high, with possibly lethal consequences.

Understanding this makes it easier for us to comprehend why so many of us seem to be hard-wired for anxiety even though, in most respects, our world is safer than ever. As LeDoux puts it, *"Bigger brains allow better plans, but for these you pay in the currency of anxiety"*.[33]

Asymmetries Induce Strong Biases

The cost of treating a stick as a snake is less, in the long run,
then the cost of treating a snake as a stick.
Joseph LeDoux[34]

Taking into account the things that bothered our ancestors in prehistoric times, we all inherited dedicated survival and reproduction problem-solving modules that are still active, despite the modern context in which we operate.

---

[33] LeDoux, J., 1998
[34] LeDoux, J., 1998

Those modules have been tailored the same way we tune fire detectors, that is, taking into account cost asymmetries.

Picture: the "paranoid" fire detector

	fire	no fire
alarm (positive P)	true positive (TP)	"false alarm" false positive (FP)
no alarm (negative N)	"miss" false negative (FN)	true negative (TN)

In hazard detection, false negatives FN (misses, i.e., there's a fire but the alarm doesn't ring) are often much more costly than false positives FP (false alarms), whence hazard detectors are often biased toward false alarms.

The Life-Dinner Principle (the Predator Runs for Dinner, the Prey for Life)

Divergence of interests between predator and prey stands as a major asymmetry at the core of the Dawkins "*life-dinner*" principle.

Animals survive risk by efficient response to alarm. For this purpose, prey and predator maintain a critical distance, corresponding to the prey's margins of safety and the predator's ability to catch its prey. Once this threshold is crossed, both prey and predator start running.

It seems though that the maximal speeds of prey (running for their lives) are systematically lower than those of predators (chasing their dinner).

Thomson's gazelle (*Eudorcas thomsonii*), among the cheetah's favorite dinner, is able to run at speeds up to 70 km/h.[35] The cheetah (*Acinonyx jubatus*), is "*one of the fastest terrestrial mammals, with reported maximum speeds ranging from 80 to 112 kilometers per hour*". From the Museum of Zoology at the University of Michigan, we learn that: *This velocity, however, cannot be maintained for more than a few hundred meters before the individual overheats. The majority of hunts end in failure*[36[*(—) Thomson's gazelles can outrun cheetahs if they can evade them for long enough because cheetahs can maintain high speeds for shorter times.*[37]

---

[35] Kingdon, J., The Kingdon Field Guide to African Mammals. San Diego, CA: Academic Press, 1997
[36] Nowak, R., Walker's mammals of the world, 6th edition. Baltimore, Maryland: Johns Hopkins University Press, 1999
[37] Extracts from the Animal Diversity Web of the University of Michigan; http://animaldiversity.ummz.umich.edu/

In this case, running for one's life does not mean escaping faster but fleeing longer. *"Slow and steady wins the race"* is the English translation of the French *"Rien ne sert de courir, il faut partir à point"*, a famous quote from Jean de La Fontaine in his fable *Le Lièvre et la Tortue* (VI, 10; *Fables*, 1668–1694).

In dangerous situations, animals protect themselves to survive. For this purpose, they have recourse to a limited repertoire of strategies, commonly known as: flight, fight, play dead. In their paper *Fear and fitness: An evolutionary analysis of anxiety disorders*,[38] psychiatrist Isaac Meyer Marks and Randolph Martin Nesse, Professor of Life Science at Arizona State University, review the evolutionary origins and functions of anxiety. They establish a refined list of how anxiety provides protection: escape (flight) or avoidance, aggressive defence, freezing or immobility, submission or appeasement.

Imagine how our ancestors might have come upon a predator in the wilderness. In a situation of acute threat, such as an encounter with a tiger, one would not have the luxury of thinking about things carefully. There would be no time to make a fine and lengthy analysis as to whether the tiger is hungry or not. In this moment, the individual faced with the threat has three options: to flee and hope to be faster than the tiger; to trust in their strength and fight, or to play dead, believing the tiger might be tricked. Unfortunately, the latter option offers experimentation primarily to the tiger, which might always decide that today's the day to try a dietary novelty, or that the best way to find out whether or not a certain body is playing dead or has really begun to decompose is to give it a good chew.

Seeing Storms Behind the Clouds

In "Seeing storms behind the clouds",[39] Andrew Galperin, Daniel M.T. Fessler, Kerri L. Johnson, and Martie G. Haselton show how our assessment of traits of anger is biased in an adaptively rational way. It is more dangerous (and costly) to underestimate another's anger (false negative error), than to suspect a happy guy of being angry (false positive).

Paranoid attitudes in human groups can lead to distrust and suspicion regarding others' motivations and intentions. In a study, Roderick Kramer,[40] a Professor of Organizational Behavior at Stanford, reported the so-called *"sinister attribution error"*. Here a student in a masters program had made an urgent phone call to a classmate the day before an exam, but the call was not

---

[38] Marks, I. M. and Nesse, R. M., 1994
[39] Galperin, A. Fessler, D. M. T., Johnson, K. L. and Haselton, M. G., 2013
[40] Kramer, R. M., 1994

returned. From a variety of reasons that might have explained why his friend had not returned the call (such as he did not get his message, or did not have time to answer or was truly busy at the time), the student inferred that he had not because the classmate did not want to speak to him.

The Coolidge Effect

Men and women display strong asymmetry in their reproductive behaviors. As psychologist Nancy Etcoff puts it in her book *Survival of the Prettiest: The Science of Beauty*, whereas the *"average woman can produce no more than eleven children, whether she has one lover or a thousand"*, a *"man can fertilize as many women as will let him because his body is constantly replenishing sperm"*.[41] Thus the bottleneck for human reproduction is female reproductive capacity. Pregnancy lasts 9 months, followed by lactation period, and female fertility is highest in a 20 year-time window. Hence "slots" for reproduction are few; therefore they enjoy high value for the female that will "allocate" them with care.

Evolutionary biologist Robert Trivers' theory of parental investment[42] states that the sex that makes the greatest investment in producing, nurturing, and protecting offspring—like women in the human species—will be more discriminating in mating: ladies are "choosy". By contrast, the sex that invests less in offspring—men—will compete to seduce the selective opposite sex; some men will restrict access to women, at the expense of other men.

Such asymmetries have behavioral consequences, as illustrated by the so-called "*Coolidge effect*":[43]

> *The following story is told about the time when President Calvin Coolidge and the first lady were being given separate tours of newly formed government farms. Upon passing the chicken coops and noticing a rooster vigorously copulating with a hen, Mrs. Coolidge inquired about how often the rooster performed this duty.*
>
> *"Dozens of times each day"* replied the guide.
> Mrs. Coolidge asked the guide to *"please mention this fact to the president."*
> When the president passed by later and was informed of the sexual vigour of the rooster, he asked, *"Always with the same hen?"*
> *"Oh, no,"* the guide replied, *"a different one each time."*
> *"Please tell that to Mrs. Coolidge,"* said the President.

---

[41] Etcoff, N., 1999
[42] Trivers, R. L., 1972
[43] Buss, D., 2003

No wonder men tend to over-interpret women's signals as sexual appeals.[44] Such a bias is surely statistically to the advantage of those men carrying it! Unfortunately it can lead to sexual harassment!

Romance, Sex, and Other Biases

A panoply of biases can be related to sex differences, as explored by David Buss, one of the main specialists in the field.[45]

Because the bottleneck for human reproduction is female capacity to give birth, men of all ages generally prefer younger women, and especially women in their most fertile phase. However, as reproduction opportunities for women are rare, women are generally attracted by men that display signs of power, because power correlates with the capacity to hold resources that will be beneficial for the success of their offspring. Though seen as provocative by some, power is a good start for romance, and such explanations make sense in the light of an evolutionary perspective on love. Spotting and focusing on resources might well be a parsimonious way in which evolution selected to assist in the identification of "Mr Right".

The Blackstone Ratio In Courtrooms

Even in matters of justice, the realm of balanced judgement, one can find asymmetry, as illustrated by the so-called "Blackstone ratio":[46] *"It is better that ten guilty persons escape than one innocent suffer"*. Is justice tuned like a fire detector? We are much more sensitive to one innocent person being wrongly considered guilty than several culprits escaping punishment. Plenty of movies tap into this human bias: *The Fugitive* (1993, by director Andrew Davis, based on the 1960s television series of the same name created by Roy Huggins), *The Wrong Man* (1956, by director Alfred Hitchcock), etc.

---

[44] Buss, D., 2003
[45] Buss, D., 2003; Buss, D., 2014
[46] Connolly, T., 1987

4
We Like Things the Way They Are

Now, we voyage across the deep blue sea where losses loom larger than gains, and cast the anchor ... You might stay there forever, paralyzed by the power of the status quo bias, nightmarish money matters, debts and tax worries, Christmas gifts, Russian roulette, or committing yourself to membership of the fitness centre.

The following two anecdotes highlight the roles of loss aversion and inertia (status quo) when our biased mind makes decisions.

> *Economist Paul Samuelson, a Nobel Prize winner, once offered a colleague the following bet:*
> *"Flip a coin. Heads you win $ 200, tails you lose $ 100.*
> *Samuelson reports that his colleague turned this bet down on the rationale that,*
> *"I won't bet because I would feel the $ 100 loss more than the $ 200 gain."*

This sentiment is the intuition behind the concept of loss aversion (although this was not the subject of Samuelson's article).[1]

Economists William Samuelson, Professor of Markets, Public Policy and Law at Boston University, and Richard Zeckhauser, Professor of Political Economy at Harvard University, relate this amusing tale about one of their colleagues:[2]

> *For 26 years, a colleague of ours chose a ham and cheese sandwich on rye at a local diner.*
> *On March 3, 1968 (a Thursday), he ordered a chicken salad sandwich on whole wheat; since then he has eaten chicken salad for lunch every working day.*

---
[1] Samuelson, P. A., 1963
[2] Samuelson, W. and Zeckhauser, R., 1988

Such an anecdote demonstrates the strength of the status quo effect, a preference for the current state of affairs. We like things the way they are!

In the previous chapter, we ended by surveying a selection of strong asymmetries in human behavior. Now, we focus on the asymmetry between our perceptions of losses and gains, where both are assessed with respect to a specific reference point, called the *"anchor"*.

Losses Loom Larger Than Gains

According to Weber's law, we respond more to variations than to absolute levels, as is the case, for instance, with differences of temperature, sound, or light.

Daniel Kahneman and Amos Tversky, two psychologists, discovered that this response to variations displays asymmetry: changes that make things worse loom larger than changes that bring about improvements. This property is one of the building blocks of Prospect Theory[3]

 A Small Dose of Prospect Theory

Daniel Kahneman and Amos Tversky developed Prospect Theory which considers the way people make choices in risky situations. Individuals tend to evaluate outcomes with respect to a specific reference point, called the "anchor". From this anchor, "losses loom larger than gains", that is, a double gain roughly offsets one loss. In addition, the *"value function"* (that turns outcome into utility) is concave in the domain of gains, and convex in the domain of losses. Daniel Kahneman later won the Nobel Prize in Economics for that theory (while Amos Tversky had since passed away).

---

[3] Kahneman, D. and Tversky, A., 1979

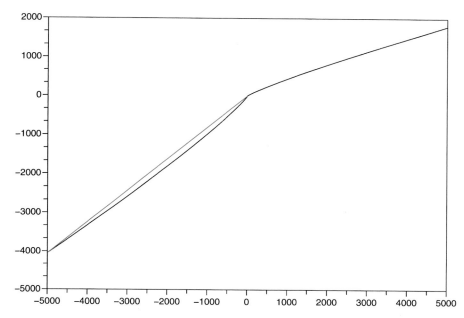

Picture: the value function according to Prospect Theory

The Prospect Theory value function above displays two regimes. To the left, in the domain of losses, it is convex. Convexity is a mathematical term to express the fact that, if you trace a straight line between two points on the curve, this line will be above the curve.

Convexity in the domain of losses implies that you prefer a single loss, say US$ 1,000, to a series of smaller losses with the same sum, say 10 times US$ 100. Let's take an example. In France, the bill in a restaurant is called the *"addition"*, the sum of money corresponding to everything you ordered. The bill is familiarly called *"la douloureuse"*, the "painful", because it hurts you and your wallet. Of course, it is paid once and for all, at the end of the meal. What convexity implies is that it would be more painful to pay the bill for a typical French meal, step by step, item by item, first after the wine, second after the *"entrée"*, third after the main course, then after the salad followed by the cheese (not to mention the pear before the cheese), and finally after the dessert, not forgetting the *"digestif"*!

To the right, in the domain of gains, the Prospect Theory value function is concave. Concavity is the property that, if you trace a straight line between two points on the curve, this line will lie entirely below the curve (the concavity on the right is less visible in the figure than the convexity on the left).

What does concavity mean in the domain of gains? Suppose you own ten Ferraris and your girlfriend offers you an eleventh one. You do not mask your joy, but you remember with nostalgia the incomparable feeling you experienced with the first one (Ferrari, of course). In other words, concavity means that an additional unit of income has more value when your income is low than when it is high.

Paying One's Taxes At Source Feels Less Painful

Money is important to us all, and yet we tend to make seriously non-rational decisions about how to deal with it. For example, people are inclined to be *"debt averse"*; they pay off mortgages and student loans quicker than they have to, even when the rate they are paying is less than what they earn on safe investments. At the same time, nobody likes paying taxes!

In some countries, like the UK, taxes are withheld directly from pay checks at source. This system is perceived as less unpleasant than a system used elsewhere, like in France, in which income taxes are paid one year after wages have been cashed. But why?

According to economist Richard Thaler[4], the least unpleasant type of loss is the reduction of a large gain. It is less painful and stressful to have to pay one's taxes at source—not to mention the fact that taxes are easier to collect in this way. Because they have never really "had" the money, British workers feel the loss of their taxes much less keenly than people who have seen the money in their bank account, as in France, and then see the new and much smaller sum that remains after all the taxes have been paid!

In both cases the amount of money received in any taxpayer's bank account plays the role of an "anchor". In the UK, the anchor is a net amount of money and there won't be any more deductions, that is, no losses. For French taxpayers, the anchor is the gross amount of money that will be subjected to future deductions, and that looks like a loss!

Don't Wrap Your Christmas Gifts In a Single Package

Have you noticed how, when someone is trying to sell you something, they often "throw in" numerous small gifts to make it seem as though you are getting a lot for your money? Car companies seem to specialise in elaborate "all in one" offers, with lengthy lists of what you are going to get for your hard-earned cash.

On the other hand, people prefer to pay once and for all than experience many small debts/losses (hence the car package including several options).[5]

---

[4] Thaler, R., 1985
[5] Thaler, R., 1980

In other words, people tend to prefer multiple small gains to one larger gain, and a large loss to multiple losses. As we have seen, this observation can be explained by the shape of the "value function" in Prospect Theory.

> **Helpful Tips On How to Please When You Offer Presents**
>
> You can impress your friends and relatives more by giving lots of small presents rather than just one large one! (Sure, you'd better refrain from telling them you used the concavity of their value function in the domain of gains…).
> You might also consider giving the best gift at the end.[6]
> (We will be looking at the so-called "peak-end" rule later.)

Mental Accounting

Richard Thaler[7] coined the term *"mental accounting"* for our propensity to open and close separate accounts. He tells an amusing story that illustrates how people tend to engage in mental accounting to deal with their household budget, providing the following anecdote:

> Mr. and Mrs. L and Mr. and Mrs. H went on a fishing trip in the northwest and caught some salmon.
> They packed the fish and sent it home on an airline, but the fish were lost in transit. They received $ 300 from the airline. The couples take the money, go out to dinner and spend $ 225. They had never spent that much at a restaurant before.[8]

For an economist, the additional budget of $ 300 should enlarge their global budget and be the starting point of a new optimization calculus. However, what happened there is that a heuristic created a behavior of the "mental accounting" type. It seemed to the couples, that it was much more gratifying to add that unexpected income to a budget subset where the spending would be more visible, rather than diluting it into the global savings.

It might be that mental accounting is a by-product of our modular mind: perhaps our mental modules perform their tasks in a loosely coordinated manner? It could also be a symptom of our social mind, prompt to reciprocate, and not always paranoid. As we will see later when we discuss tit-for-tat,

---
[6] Do A.M., Rupert A. V. and Wolford G., 2008
[7] Thaler, R., 1985
[8] Thaler, R., 1985

if your colleague Jim offers you a coffee, your mind starts to open the "Jim paid me a coffee" account, so that you feel obliged to return the favor.

We Are Biased Towards the Status Quo

Economists William Samuelson and Richard Zeckhauser coined the term "status quo bias" to describe our preference for the current state of affairs—we are apparently willing to keep existing situations in our daily lives even when they do not satisfy us. Who does not adjust to a drafty room rather than getting the window fixed, or put up with a cracked monitor because replacing it seems like a hassle?

Economist John Kenneth Galbraith described an extremely common reaction to change: faced with the choice between changing one's mind and proving that there is no need to do so, almost everyone gets busy on the proof. Imperfect as things may be, at least we are familiar with the current situation and know how to operate in it.

Suppose You Are Compelled to Play Russian Roulette

Economists Richard Zeckhauser and W. Kip Viscusi[9] imagine the following situation:

> Suppose you are compelled to play Russian roulette, but are given the opportunity to purchase the removal of one bullet from the loaded gun.
> Would you pay as much to reduce the number of bullets from four to three, as you would to reduce the number of bullets from one to zero?

Would you not be willing to pay much more for a reduction of the probability of death from 1/6 to zero than for a reduction from 4/6 to 3/6? Of course! This is an illustration of the so-called "certainty effect"[10]: we prefer the certain outcome, no bullet, where all risk has vanished. Surprisingly, economists calculate we should value the latter choice more, were we to be so-called *"expected-utility maximizers"* in the economic jargon (but in many circumstances we do not seem to follow the mainstream economic theory on decision-making under risk).

There is even a fictitious scene, shot by Michael Cimino in *The Deer Hunter* movie, in which the main character played by Robert De Niro asks his guards

---
[9] Zeckhauser, R. and Viscusi, W. K., 2000
[10] Tversky, A. and Kahneman, D., 1986

to add more bullets to his gun during a terrifying scene of Russian roulette, balancing the frequency at which he will shoot a bullet to a guard or to himself. When the number of bullets increases, there are six possible shots and a specific number of guards to be shot!

Losses and gains take their full meaning relative to a reference value called the "anchor". Here is an intriguing test that will give you a feel for the power of anchoring.

> **An Amusing Test of Elementary Arithmetic**
> $8 \times 7 \times 6 \times 5 \times 4 \times 3 \times 2 \times 1 = 1 \times 2 \times 3 \times 4 \times 5 \times 6 \times 7 \times 8$?
> Ask one of your friends to mentally calculate the following in 5 s:
> $8 \times 7 \times 6 \times 5 \times 4 \times 3 \times 2 \times 1$
> Now write down the estimate.
> Done? Then ask somebody else to mentally compute the following in 5 s:
> $1 \times 2 \times 3 \times 4 \times 5 \times 6 \times 7 \times 8$
> Now write down the estimate.

The outcome from the second exercise will probably be lower than the first. This famous experiment from Kahneman and Tversky[11] gave the following outcomes:

$8 \times 7 \times 6 \times 5 \times 4 \times 3 \times 2 \times 1$: median 2,250
$1 \times 2 \times 3 \times 4 \times 5 \times 6 \times 7 \times 8$: median 512[12]

(Median 2,250 means that half of the subjects are below or above that figure.)

People do not have enough time to perform all the multiplications, so they will stop their calculations after a few steps. When calculating $8 \times 7 \times 6 \times 5 \times 4 \times 3 \times 2 \times 1$, people start by multiplying 8 by 7, 56 by 6, 336 by 5, etc., anchoring the expected outcome in high numbers (8, 56, 336, etc.). In contrast, starting the multiplication by the lowest number 1 gives 1, 2, 6, etc., leading to a low expected result.

Hands Off My Stuff!

Daniel Kahneman, Jack L. Knetsch, and Richard H. Thaler[13] conducted a simple exercise involving the exchange of two goods. One group of students was

---

[11] Kahneman, D. and Tversky, A., 1974
[12] The correct answer is that $8 \times 7 \times 6 \times 5 \times 4 \times 3 \times 2 \times 1 = 1 \times 2 \times 3 \times 4 \times 5 \times 6 \times 7 \times 8 = 40,320$
[13] Kahneman, D., Knetsch, J. K. and Thaler, R., 1991

given a coffee mug and asked to describe it. The other group was initially given a candy bar, together with some information about it. The experimenter offered the first group the opportunity to change their mug for a candy bar. Similarly, the same type of coffee mug was offered to the second group instead of their candy bar. In both groups, the mugs and candy bars were exactly the same. When given a simple choice between mug and candy bar, 56% of the students preferred the mug. Moreover, only 10% accepted the offer of exchanging their candy bar for a mug when the candy had been given to them first, while 89% of those given a mug initially turned down the offer of candy instead.

This response is deeply rooted in principles of law and justice, as in the case of the old saying that *"possession is nine tenths of the law"*. The tendency to prefer to keep something we already have rather than getting something else is a bias known as the *"endowment effect"*.[14]

Coca Cola Fans Leaned for the Old Coke Classic

In 1986, during a typical hot summer when Americans enjoy cooling down with one of their favourite soft drinks, the Coca Cola Company launched a new version of their famous drink, with a sweeter taste. The new product was named "Coca Cola", just like the version it was replacing. But, as outlined by Samuelson and Zeckhauser[15], the Coca Cola Company failed to recognize the importance of the status quo bias. In blind taste tests, consumers (including loyal Coke drinkers) were found to prefer the sweeter taste of new Coke over old by a large margin.

However, informed consumer preferences may differ from blind preferences. In other words the company did not quite anticipate consumer reactions when they became fully aware of the name of the recipe they were tasting (Classic versus New). Coke drinkers' loyalty to the status quo (Coke Classic currently outsold new Coke by three to one) far outweighed the taste distinctions recorded in blind taste tests. This status quo bias applied only in the US, where consumers had been accustomed to the brand and had remained loyal for decades. The Coca Cola company president appeared in a TV advert to promise to return post haste to the older version.

An amusing marketing side story is that the 1985 mistake was soon replaced by a new opportunity to segment the Coca Cola offer! First, instead of cancelling the new version of the soft drink, the Company maintained two versions of the drink, named Classic Coke and New Coke. This was an attempt to please the loyal consumer with Classic and recruit new consumers in search of smoothness and greater sweetness, namely Pepsi Coke type consum-

---
[14] Thaler, R., 1980
[15] Samuelson, W. and Zeckhauser, R., 1988

ers, with New Coke. This particular segmentation was only maintained for a while, however, and then the Company went back to the status quo ante, with only one Coke (slightly different though regarding the vanilla flavour), before adding new flavours to the original recipe (lemon, cherry…) or various kinds of sweetness (Diet, Light, Zero…).

A Small Town In Germany

Samuelson and Zeckhauser[16] relate another example of a striking anchoring process. Some years ago, the West German government undertook a strip-mining project that required the relocation of a small town underlain by lignite. At their own expense, the government offered to relocate the town in a similar valley nearby. Specialists suggested scores of town planning options, but the townspeople selected a plan extraordinarily like the serpentine layout of the old town; a layout that had certainly evolved over centuries without any organised planning.

The Drubeck Brothers' Story

Robert Cialdini[17], Professor Emeritus of Psychology and Marketing at Arizona State University, reports a story by Leo Rosten, about the Drubeck brothers, Sid and Harry, who owned a men's tailor shop in the 1930s.

> *Whenever the salesman, Sid, had a new customer trying on suits in front of the shop's three-sided mirror, he would admit to a hearing problem, and, as they talked, he would repeatedly request that the man speak more loudly to him. Once the customer had found a suit he liked and had asked for the price, Sid would call to his brother, the head tailor, at the back of the room,*
> *"Harry, how much for this suit?"*
> *Looking up from his work—and greatly exaggerating the suit's true price—Harry would call back,*
> *"For that beautiful all-wool suit, forty-two dollars."*
> *Pretending not to have heard and cupping his hand to his ear, Sid would ask again. Once more Harry would reply,*
> *"Forty-two dollars."*
> *At this point, Sid would turn to the customer and report,*
> *"He says twenty-two dollars."*
> *Many a man would hurry to buy the suit and scramble out of the shop with his "expensive = good" bargain before Poor Sid discovered the "mistake."*

---

[16] Samuelson, W. and Zeckhauser, R., 1988
[17] Cialdini, R. B., 1984

In this anecdote, the customer was provided with $ 42 as a reference price for the suit, to which he compared the $ 22 price. Because of this, the customer felt as though he'd just made himself a profit of $ 20.

"Tell Me Something I Didn't Learn In Hotel School"

Economist Daniel McFadden, Nobel Prize winner, has a nice anecdote in his paper *Free Markets and Fettered Consumers:*[18]

> A study by Itamar Simonson and Amos Tversky (1992) finds that when products are positioned so that one appears to be a bargain, a form of anchoring, then consumers will flock to the apparent bargain alternative.
> When I told a friend who owns a Boston seafood restaurant that he could use this result to reposition his wine list and increase his profits, his response was "tell me something I didn't learn in hotel school."

Still, goods on special offer, showing the former price beside the new price on a supermarket shelf, are everyday examples of such anchoring tricks.

Picture: a wine under promotion in a British supermarket

The crossed-out price of £ 9.99 acts as a reminder of the former price anchor, whereas the new £ 6.49 appears to be more salient and acts as the new price anchor. The result is expressed as a gain (a bargain) of £ 3.50.

---
[18] McFadden, D., 2006

It Took 400 Years Before We Used Lemon Juice to Avoid Scurvy

In her paper *"Bridging the gap between scientists and practitioners: The challenge before us"*,[19] Linda Carter Sobell—Professor of Clinical Training at Nova Southeastern University, Florida—describes the problem of getting practitioners to accept the use of evidence-based health care. She begins her discussion by reminding us how difficult it was to convince the British Navy to accept the use of lemon juice by its sailors to avoid scurvy and, ultimately, death in 62% of cases. Apparently some 400 years went by between the discovery that lemon juice could play a useful role here and its final acceptance as part of the sailor's regular diet.

When We Anchor Our Assessment On Mere Fortune

Kahneman and Tversky[20] designed a simple experiment to illustrate the strength of anchoring in people's assessment. In that experiment, there were two groups playing at the wheel of fortune. In each group, one player was spinning the wheel of fortune in front of the others. Each time the wheel was turned, the resulting number was shown to the group. The first group of players was asked, "Do you think the percentage of African countries in the UN is above or below this number given by the wheel?" (Although, of course, the question was in no way related to the figure given by the wheel.) The median estimate of subjects who saw the wheel marking 65 was 45% (meaning that 45% of the players thought that there were more than 65 African countries in the UN). The median estimate of subjects (from a second group of players), who saw the mark 10 was 25% (meaning that 25% of the players thought that there were more than 10 African countries in the UN). In this experiment, the mark of the wheel acted as an anchor to judge the number of African countries, although the two results were completely unrelated.

It seems from this experiment that, in ignorance, we assess from any foothold, even one delivered by a wheel of fortune. It is as if the biased mind preferred to grasp the only figure available, even though it may not be relevant, rather than make an assessment from scratch.

---

[19] Sobell, L. C., 1996
[20] Kahneman, D. and Tversky, A., 1974

Did Gandhi Live More Than 140 Years?

In a similar study by Fritz Strack and Thomas Mussweiller[21], subjects were asked how old Gandhi was when he died. Answers were influenced by an initial comparison with an irrelevant, indeed absurd, anchor. If people were first asked whether Gandhi died before or after the age of 140 and then to estimate how long he lived, they guessed an average of 67 years. However, if they were asked first if he died before or after the age of 9, on average they guessed that he had lived for 50 years.

Assuming that the subjects of this experiment did not know Gandhi's age at death, we can see that, when presented with figures such as 9 or 140, however absurd they may seem at first glance, people inferred that the experimenter was giving them a clue as to his longevity, suggesting that Gandhi might have died when he was either very young or very old. In the absence of any other relevant information, the mental shortcut provided by the anchor acts as *the key* piece of information to guide judgment.

Maybe the subjects were expecting a cooperative attitude from the experimenter, and thought that he was giving them a valuable cue? We'll come back to this issue when we discuss cooperation.

When Less Is More

Depending on the anchoring, less can look like more. Better to be the best of (cheap) scarves than the worst of (expensive) coats, mimicking the Argentinian saying that *"Más vale ser cabeza de ratón que cola de león"* (better to be a mouse's head than a lion's tail).

Christopher K. Hsee[22] conducted research to test the "less is better" hypothesis. In one study, participants were asked to evaluate and compare two possible gifts: a woollen coat at $ 55 and a scarf at $ 45. In that general store, prices of woollen coats range from $ 50 to 500, whereas woollen scarves can be had for between $ 5 and $ 50.

Considering the respective values of the two items ($ 45 versus $ 55), one would expect participants to opt for the gift of the coat. However, they perceived the gift of a $ 45 scarf as more generous than the gift of a $ 55 coat.

In the present case, at a price of $ 55 the coat was judged to be quite cheap, bearing in mind that coats can sell at up to $ 500. Conversely, one could perceive the woollen scarf as expensive at $ 45, especially when they sell at $ 50 at the very most.

---

[21] Strack, F. and Mussweiler, T., 1997
[22] Hsee, C. K., 1998

What about you? Don't you feel happier when your ice-cream cone is overfull? Aren't you more confident as a speaker in a small but crowded conference room, as opposed to a large, relatively empty one?

We Succumb Over and Over Again to Committing Ourselves

See if you recognize yourself in the familiar examples of the bias called "psychological commitment".

Thaler[23] provides several examples. Anyone who has prepaid for a concert or theatre series feels she *must* attend each concert or play, regardless of how little she feels like it, or even if she is running a fever. She's already paid, so if she does not go, she'll feel positively cheated out of an evening's entertainment. Would she feel the same way if the tickets had been given to her for free? Guess not! Similarly, in order to recover his annual membership fee, a sports enthusiast continues to play tennis three times a week, despite a painful tennis elbow.[24]

A similar perspective can be found in numerous examples from the business world. For example, Lockheed continued to build the unprofitable L-101 aircraft (with the aid of Congressional funds) in the vain hope of recovering its past investment. The day after it finally conceded defeat and cancelled it, Lockheed shares rose 18%. Such wrong-headedness is known as the *"Concorde Fallacy"* because of a similar story involving the iconic French and British plane.[25]

Coffee shops and restaurants that issue loyalty cards promising a free cup of coffee or snack for every ten purchased are manipulating their consumers by playing on the commitment bias. As well as offering a clear incentive to continue, in the form of the free drink or snack, possession of the card can give the holder a sense of commitment to the outlet; a feeling of being "in a relationship" with it rather than with other stores offering similar wares.

Fitness centres that sell yearly memberships are doing the same thing. We take out membership for a year with the idea that we will make a commitment to go to the gym on a regular basis. People are encouraged to do so by clubs offering lower monthly rates on an annual basis. The feeling is that they would be less likely to go if they had to pay each time. But guess what? The reality is that many of the people who subscribe to a gym for a year, paying

---

[23] Thaler, R., 1980
[24] Samuelson, W. and Zeckhauser, R., 1988
[25] Samuelson, W. and Zeckhauser, R., 1988

a hefty sum upfront, still rarely go, and some do not even go once, although they hoped that paying for the year would be a strong enough incentive! We are all multiple selves, and all too often the self that would rather stay home, watch TV, and eat potato chips is the one that wins the toss.

Steve Martin, who co-wrote a book with Noah Goldstein and Robert Cialdini[26], reported in the *Guardian* newspaper on a restaurant-based study showing that waiters received many more tips by giving customers a mint just before they put the bill down. And if they gave two mints, they were not only more likely to get a tip, but it would probably be bigger. The success of this ploy depended on the waiters giving the mints *before* presenting the bill; putting the mints *down after* the bill didn't work! This is another example of reciprocity techniques aimed at building commitment. By giving the customer the mint, they were establishing a norm of giving.

So how is this information useful to the man or woman in the street? In many ways! In our personal lives, for example, how many of us have persisted with a romantic relationship that has long exceeded its sell-by date, simply because of a feeling of having made a commitment to the other person, even though neither party to the relation is happy.

Desperately Seeking…Confirmation

Everything which has occurred since then has served to confirm my original supposition, and, indeed, was the logical sequence of it.
Arthur Conan Doyle, *A Study in Scarlet*, 1887

English psychologist Peter Wason[27] reminds us that *"scientific inferences are based on the principle of eliminating hypotheses, while provisionally accepting only those which remain"*. He is following Karl Popper, the famous philosopher of science. As Popper stressed that empirical scientific theories should be open to the possibility of falsification and refutation, Wason wondered if our minds were more open to refute or to support hypotheses. The title of his 1960 paper, *"On the failure to eliminate hypotheses in a conceptual task"*, points to the existence of a *"confirmation bias"* that leads people to misinterpret new information as supporting previously held hypotheses.

Inducing overconfidence, such a bias may also lead people to believe strongly in a false hypothesis despite receiving a huge amount of information that goes against it. Maybe our cognitive limitations are at work here. Joseph

---
[26] Goldstein, N. J., Martin, S. J. and Cialdini, R., 2009
[27] Wason, P. C., 1960

LeDoux suggests that the information stored in our memory, once *"activated and made available to working memory"*, influences lower level information-seeking processers that scrutinize the environment with a bias.[28]

When Karl Popper Caught Psychiatrist Alfred Adler In Confirmatory Bias

Economists Matthew Rabin and Joel L. Schrag[29] report an anecdote by Karl Popper. In the lecture *Science: Conjectures and Refutations*[30]—given at Peterhouse, Cambridge, in Summer 1953, and organized by the British Council—Popper evokes those of his friends who were admirers of Marx, Freud, and Adler and who were impressed by the explanatory power of their theories. He stresses that they saw *"confirming instances everywhere"*, and he uses a personal anecdote to expose the process of *"verification"* by which they strengthened their conviction over time.

> *As for Adler, I was much impressed by a personal experience. Once, in 1919, I reported to him a case which to me did not seem particularly Adlerian, but which he found no difficulty in analyzing in terms of his theory of inferiority feelings, although he had not even seen the child. Slightly shocked, I asked him how he could be so sure. "Because of my thousand fold experience," he replied;*
> *Whereupon I could not help saying:*
> *"And with this new case, I suppose, your experience has become thousand-and-one-fold."*

Popper goes on to express his view that *"It is easy to obtain confirmations, or verifications, for nearly every theory—if we look for confirmations"*. In his view, the weakness of Freud and Adler's theories resided in their capacity to interpret every conceivable case. In contrast, Popper makes the point that a strong theory *"forbids certain things to happen"*. Without Einstein's gravitational theory, one would expect light rays to follow a straight line when passing close to the sun. But Einstein arrived at the conclusion that light should be attracted by the massive sun, and this was subsequently corroborated by observation. Thus, Einstein's theory *"forbids"* light from going straight in the neighbourhood of a massive body. Such a theory, which actually excludes certain things, contrasts starkly with Freud and Adler's, which are so vague and flexible that it is hard to conceive of any case that is not actually compatible with them.

---

[28] LeDoux, J., 1998
[29] Rabin, M. and Schrag, J. L., 1999
[30] Popper, K., 2002

How to Influence the Perceptions of Fidel Castro

Edward E. Jones and Victor Harris[31] developed the hypothesis that, when people were free to choose how to behave, their behaviors would be attributed by others to their personal disposition, whereas the ones that looked as though they were driven by more random factors would be attributed to the situation at hand.

In their study, research subjects were asked to read essays either in favour of or against Fidel Castro. When they believed that the writers were free to choose their positions, they rated those in favour of Castro as having a very positive attitude towards him. However, when they were told that the writer's stance in the essay had been determined by the toss of a coin, while they could see the constraints created by the situation, they still found it hard not to assume that the writer had some positive feelings for Castro.

Love Is Blind, But the Neighbours Aren't

To look at the issue of personal lives and love, who has not experienced confirmatory bias in the context of unrequited love, either as the subject or object of it? The strength of our feelings colours the way we see and interpret reality. Imagine the young man who has fallen for the university librarian he sees every day. She has no idea that he exists, but he has become convinced that she has noticed him and likes him too. When he was let off a fine for a book that was a day late, he was utterly convinced that this was a breakthrough in their relationship! Surely, he thought breathlessly, this was a romantic gesture on her part, intended to make him aware of her feelings!

According to a Mexican saying, *"El amor es ciego, pero los vecinos no"* (Love is blind, but neighbours aren't).[32]

We started with the *Biased Mind* assessing losses and gains, and anchoring. Now we end by observing how our mind seeks confirmation, selectively "cherry-picking" the confirmatory information from the environment.

---

[31] Jones, E. E. and Harris, V. A., 1967
[32] Ledoux, J., 1998

5
Our Detective Mind Grasps Clues and Narrates

Transported by magic, we land in the country of illusions and superstitions: homeopathy, conspiracy, illusion of control, bad luck. Mystified, we overshadow relevant facts just to narrate a compelling story and never cease to wonder "what if...?" Along the road, we will encounter bronze medalists, regretful parents, cold readers, control freaks, poor alibis, and weak stories.

In *Craps and Magic*,[1] James M. Henslin, Professor Emeritus of Sociology at Southern Illinois University, Edwardsville, reports a scene at the crap table:

> *after one throw, the result of which was a high number, the shooter said:*
>
> *Shot too hard that time.*

It is believed that "a hard throw produces a large number, and a soft or easy throw produces a low number".

Extract from CSI Las Vegas Season 8 episode 3: "Go to Hell" script:

> *"Grissom, do you believe in a separate, living evil?"*
> *Crime scene investigator Gil Grissom answers:*
> *"You're primitive man on the savannah.*
> *You see something move out of the corner of your eye.*
> *You assume it's a hyena.*
> *You run, you live.*
> *If you assume it's the wind and you're wrong, you die.*
> *We have the genes of the ones who ran.*
> *We're genetically hardwired to believe living forces that we cannot see".*

Adaptation has a long reach and scholars uphold the idea that narration and story-telling have an adaptive value. Indeed, by imagining possible worlds, scenarios—in quiet moments—the mind could be preparing itself to react

---
[1] Henslin, J. M., 1967

appropriately to real situations and threats. Brian Boyd,[2] Professor in the Department of English at the University of Auckland in New Zealand, suggests that art and storytelling derive from play. In play—to be found in sport, games, dance, etc.—people learn to coordinate within a group and prepare themselves for real contexts where survival and reproduction are at stake. When children play hide-and-seek, they develop skills that may prove useful if they ever need to escape hostile conspecifics. When playing team sports, or when dancing, participants learn to coordinate spatially and with others, and build up aptitudes that may help them to gain the upper hand in a non-play context.

Cherry-Picking and Connecting the Dots

And when ambiguous and abundant information is fed to the biased mind, it enters into a selective search for clues, called "cherry-picking".

The Barnum Effect ("We've Got Something for Everyone")

One of the authors of this book happened to read the horoscope for Taurus (his birthday is on 14 May). What he read was troubling, as the mood described for all Taurus people was pretty close to how he was feeling at that time and could have been written for him in person:

> You need to get out and have fun and with today's infusion of positive energy, it's hard for you to do otherwise! Blow off some steam at your favourite night spot or catch that hot new movie.

The Forer Effect—also known as the "Barnum Effect" after circus and freak show owner P. T. Barnum's supposed observation that *"we've got something for everyone"*—refers to the fact that people tend to subscribe to descriptions of their personality that are supposed to have been tailored specifically for them, but are in reality vague and general enough to apply to everyone.

In 1949, psychologist Bertram R. Forer[3] gave a personality test to his students. Each student answered the questions. Afterwards, he told them they were each receiving a unique personality analysis that was based on the test results. He then asked them to rate their analysis depending on how accurately it applied to them. In reality, each received the exact same analysis, which read as follows:

---

[2] Boyd, B., 2009
[3] Forer, B. R., 1949

You have a great need for other people to like and admire you. You have a tendency to be critical of yourself. You have a great deal of unused capacity which you have not turned to your advantage. While you have some personality weaknesses, you are generally able to compensate for them. Disciplined and self-controlled outside, you tend to be worrisome and insecure inside.

At times you have serious doubts as to whether you have made the right decision or done the right thing. You prefer a certain amount of change and variety and become dissatisfied when hemmed in by restrictions and limitations. You pride yourself as an independent thinker and do not accept others' statements without satisfactory proof. You have found it unwise to be too frank in revealing yourself to others. At times you are extroverted, affable, sociable, while at other times you are introverted, wary, reserved. Some of your aspirations tend to be pretty unrealistic. Security is one of your major goals in life.

On average, the rating measuring affinity between the profile and the reader's personal life was very high at 4.26 on a scale from 0 (very poor) to 5 (excellent). Thus, the students found that the above description was pretty close to their personality. Only after the ratings were handed in, was it revealed that each student had received identical copies assembled by Forer from various horoscopes. Though all were given identical material, each one sought out the unique and exclusive. The profile designed by Forer simply brings together a number of statements that could apply equally to anyone.

The Forer effect might be linked to "self-deception", a trait favored by evolution, according to biologist Robert Trivers, at the Department of Anthropology, Rutgers University. Indeed, species practice deceit for the sake of survival and reproduction: butterflies display eyespots on their wings that mimic the eyes of other species (deceiving predators), orchids mimic female bees and wasps (increasing pollination), girls use make-up, and boys wear jackets with wide shoulders. Trivers argues that humans deceive themselves, trusting something that is not true as a better way to convince others of that truth.

Cold Readers Can Mystify Strangers by Knowing All About Them!

The Forer effect can be put to good use by so-called psychics and mediums. Whether through design or instinctively, they tap into their clients' tendency to recognize themselves in descriptions of character and personality. Then, they can wow them by appearing to "know so much", even though they have never seen them before. This approach could also be wisely employed by the would-be seducer who seeks to entrap the target of his attraction with a penetrating insight into what makes her tick, or by the cold-caller in search of new sales prospects and finely attuned to the way people behave over the phone.

Such propensity or bias can be easily exploited by so-called *"cold readers"*. *"Cold reading"* is a set of techniques that allow a *"mind reader"* to give the impression that he knows much more about someone than he actually does. On the one hand, every clue (age, body language, clothing, tone of voice) can offer insights to the mind-reader. On the other hand, the reader can deliver general statements, fit for anyone, which the individual takes for himself. By *"cherry-picking"* from the general statement, the individual unwittingly provides new clues.

In his paper *"Cold Reading: How to Convince Strangers That You Know All About Them"*,[4] Ray Hyman—Professor Emeritus of Psychology at the University of Oregon in Eugene, Oregon—shows how pitchmen, encyclopaedia salesmen, hypnotists, advertising experts, evangelists, confidence tricksters, and fortune tellers use this widespread human tendency for "cherry-picking". He provides a classic example, originally written for female college students in the 1950s:

> *You appear to be a cheerful, well-balanced person. You may have some alternation of happy and unhappy moods, but they are not extreme now. You have few or no problems with your health. You are sociable and mix well with others. You are adaptable to social situations. You tend to be adventurous. Your interests are wide. You are fairly self-confident and usually think clearly.*

It appeared that most subjects presented with this portrait found it to be a fairly accurate description of them. Written more than 60 years ago, such descriptions still echo in us, revealing continuity in our nature, despite the upheavals of the modern world.

How Venusian Artists "Cold Read" Female Targets

So-called *"Venusian artists"*[5] claim to specialize in the art of seducing women. They subtly tailor cold reading to small talk between men and women. Let's appreciate the following examples of cleverly crafted scripts, picked up here and there.

As a girl, how would you react if a boy was telling you *"that he felt that you have a secret talent that you have never used in your favour?"* or *"You have a considerable capacity within, that you did not know how to use till now"*.

Or if somebody tells you that *"Your relationships with certain people in your past have caused you many problems?"* Would you not agree?

---

[4] Hyman, R., 1977
[5] www.venusian artists.com

What about: "*I see you as an extrovert person, nice externally, but I sense a fragile girl within… you hide feelings so deeply that not even your friends know that they exist.*"

Or: "*I see that you are very friendly, that friendship comes easily to you, but I guess that, on some occasions, you feel lonely.*"

Or: "*Others see your outside personality as representing someone self-disciplined and controlled. However, within, you tend to be insecure and to worry.*"

Or: "*Sometimes you have had serious doubts as to whether or not you made the correct decision or did the right thing.*"

Or: "*You're a girl who likes to call attention to herself, possibly because sometimes you feel alone or misunderstood. I know that you seem to be very friendly, flirtatious and affectionate. Inside, however, you are a bit shy*".

Heuristics and Mental Shortcuts Fill the Dots

When information is scarce, the biased mind fills the empty slots in between the few "dots" of available data. Our mind resorts to tricks, mental shortcuts, and heuristics—what Gerd Gigerenzer calls the "*fast and frugal rules*"—a process that turns up in narration, story-telling, illusion, and magical thinking.

Judgments made in difficult circumstances can be based on a limited number of simple, rapidly-arrived-at rules ("*heuristics*"), rather than formal, extensive algorithmic calculus and programs. Often, even complex problems can be solved quickly and accurately using such "*quick and dirty*" heuristics. However, equally often, such heuristics can be beset by systematic errors or biases.

Mental Shortcuts Ground Optical Illusions

In his paper *I Think Therefore I Err*, Gerd Gigerenzer shows an example of an optical illusion and the following explanation by a mental shortcut based on an adaptive perception.[6]

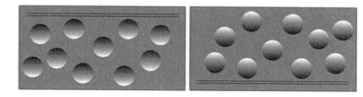

Picture: an optical illusion (from Gigerenzer)

---
[6] Gigerenzer, G., 2005

Light comes from above! The dots on the left-hand side of the image above appear to be holes, while the dots on the right side appear to be "sticking out", although neither side uses three dimensional holograms.

For Gigerenzer, this perceptual illusion helps to explain how our mind functions: our brain uses simple adaptive mental shortcuts based upon our environment to draw conclusions about the nature of the dots.

After thousands and thousands of generations of humans on Earth, our mind recalls that natural light comes from above us, and that there is only one source of light, the sun. If shade covers the upper part, dots look like holes; if shade invades the lower part, then dots look like bumps.

As Steven Pinker puts it, *"our eyes and our circuits for vision are highly elaborate devices, improved generation after generation, for the benefit of survival and reproduction."*[7] It is so natural for us to open our eyes and see the world around us that we miss the elaborate processing of visual stimuli. As the saying goes, *"we do not see the world as it is, the world is as we see it."*

The famous checker-shadow illusion (figure below), by Edward H. Adelson, Professor of Vision Science at Massachusetts Institute of Technology, is another example of how our mind can be fooled by our adaptive skills.

Picture: the Adelson checker illusion (©1995, Edward H. Adelson)

Together, our eyes and brain try to determine the colours of the object, but just measuring the light coming from a surface (the luminance) is not enough: a cast shadow will dim a surface, so that a white surface in shadow may be reflecting less light than a black surface in full light. The visual system uses several tricks to determine where the shadows are and how to compensate for them, in order to determine the shade of grey "paint" that belongs to the surface. The image on the right hand side, joining squares A and B with two vertical strips coloured the same shade of grey, makes it more obvious that

---

[7] Pinker, S., 1997

the two squares are the same colour (you may spot this by moving your eyes quickly).

In shadow or not, a check that is lighter than its neighbouring checks is probably lighter than average, and vice versa. In the image, the light check in shadow is surrounded by darker checks. Thus, even though the check is physically dark, it is light when compared to its neighbours. Conversely, the dark checks outside the shadow are surrounded by lighter checks, so they look dark by comparison.

On many occasions our mind looks for clues in the context, as in the following stunning anecdote.

"Watch the Borders" Wrote Former FBI Director Edgar J. Hoover

Edgar J. Hoover was the first Director of the Federal Bureau of Investigation (FBI) of the United States. In *The Hoover Legacy, 40 Years After, Part 5: A Day in the Life*, on the official website of the FBI, one reads that Hoover liked to write, with blue ink, on the margins of memos[8]. One day, his staff received an annotated memo with the warning *"Watch the borders"*

In the book, *No Left Turns: The FBI in Peace and War*,[9], former agent Joseph L. Schott portrays Hoover as a rigid man who terrified everyone: "Nobody asks him to explain anything. You just say, 'Yes sir', and then try to figure out what he wants." Therefore, nobody dared to ask him why borders should be watched. The FBI website reports: *The "watch the borders" command briefly set off a flurry of concern, as it was misinterpreted to mean that there was some issue involving our national boundaries with Canada or Mexico.* In fact, the message related to the borders… of the memo paper: a young special agent had tried to cram too much material onto a page, and Hoover was warning the writer of the memo to leave room in the borders.

The agents interpreted the word *"borders"* as *"frontiers"* because they knew of the authoritative personality of Edgar J. Hoover, who was obsessed with national security. That was the context. Combined with the content of that memo, it led them to an incorrect interpretation of Hoover's instruction.

Cooperative Efforts In Conversation

Herbert Paul Grice's conversation maxims remind us that some untold element may be implicit, letting the listener grasp clues from the context. For instance, if we just tell you,

---

[8] https://www.fbi.gov/news/stories/2012/september/the-hoover-legacy-40-years-after-part-5
[9] Schott, J. L., 1975

> John began to clean his room,

you deduce that he did it. Indeed if he hadn't, you would have expected the sentence to be:

> John began to clean his room, but he stopped after a while.

In the same vein, if you're told:

> I went to the cinema yesterday,

then you'll understand that I went to the cinema, and also that I watched the movie, that is, I didn't just stand in the lobby.

Grice[10] saw a general principle of cooperative efforts in conversation, assuming that our mind reaches out, beyond enunciated words, for untold cooperative cues. As Steven Pinker says, *"sentences in a spoken language like English or Japanese are designed for vocal communication between impatient, intelligent social beings. They achieve brevity by leaving out any information that the listener can mentally fill in from the context."*[11] So the cooperative assumption of Paul Grice can be grounded in evolutionary principles that cooperation boosted human reproductive success on the planet, and that our minds are designed for that purpose. When Joan speaks to Peter, she conveys her thoughts through the linear channel of a sequence of words. She does this in an economical way, mobilizing the abilities of Peter's mind to "decode" the signal sequence and turn it into thoughts.

What Makes a Good Alibi?

Thinking about exceptional events—for which there are no relevant scenarios in the past that help one to judge how likely they are to take place—we often build scenarios that lead from the present situation to that unique event. The plausibility of the possible stories that come to mind gives us a clue as to the likelihood of that event. If we are unable to come up with a plausible scenario, the event seems highly unlikely. The cognitive limitations of our mind mean that, considering the numerous variables or interactions that may have produced a specific event, only the most easily available and simple stories are considered to be likely.

---

[10] Grice, H. P., 1975
[11] Pinker, S., 1997

In courtroom arguments, credibility is often determined depending on how well the prosecution or defence puts forward the evidence. Story-telling is important here, as the relative consistency of various arguments can have a big impact on whether or not people consider the scenario to be probable.

What makes an alibi weak or strong: truth or credibility? In their paper *What makes a good alibi? A proposed taxonomy*,[12] Elizabeth A. Olson, University of Wisconsin, and Gary L. Wells, Professor in the Department of Psychology at Iowa State University, bring to light the distance that stands between true innocence and the quality of the alibi. They examined the cases of individuals convicted by juries on the basis of eyewitness identification testimony, but whose innocence was later established thanks to forensic DNA testing. Chillingly enough, they point out that prosecutors often exploited "weak alibis" as evidence for the (supposed) guilt of these innocent people.

Therefore, "proving" an alibi is not that easy, even when you are innocent!

According to Olson and Wells, "*the mere labelling of a statement as being an alibi evokes a sense of disbelief…*" Compare the straightforward and neutral question: "*Where were you last night?*" with the more inquisitorial: "*What is your alibi for where you were last night?*" In the latter case, an answer will seem to the prosecutor to be a form of denial of criminality, when at the same time such denials are often perceived by sceptical prosecutors as incriminating!

Denials such as "*I was not there*" as opposed to positive statements such as "*I was in the shop that evening*" tend to be interpreted as dubious. Positive statements appear to be more sincere.

 A Helpful Tip If You Are Not Guilty, Oops…If You Are Innocent

Communication specialists know that a positive sentence, such as "*I am innocent*" is more convincing that a negative formulation, "*I am not guilty*". It seems that our biased mind discards the negative "*not*" and only retains the word "*guilty*".

We think more naturally in terms of narratives, anecdotes, and stereotypes than theoretical constructs. Plot theories and brand story-telling are common illustrations of the power of narration. This is something long known to advertisers and there are many examples of storytelling in advertising, like the slogan "*You've come a long way, baby*" for Virginia Slims cigarettes. This linked

---

[12] Olson, M. and Wells, J. L., 2004

a brand of cigarettes with the narrative of women's liberation, which is easily called to mind by potential female consumers, the idea being that they would feel themselves "liberated" by smoking a particular brand of cigarettes—a highly controversial promise considering the addiction.

Another example, again for cigarettes, is provided by the well-known Marlboro Man advertisements, which tap in to the well-worn narrative of the rugged, husky cowboy who just does his own thing.

Good stories are easy to remember, which is why urban legends such as the "one about the escaped madman with the hook for a hand" and "the one about the ghost hitchhiker" are told over and over again. Such stories draw the listener in quickly, get the narrative across, and relay a clear and unambiguous conclusion.

What If?

Scenarios are sequences of events related by causal links. What if we changed one event? Joseph LeDoux points out that imagining alternative scenarios goes with the expansion of our brain (and is paid "*in the currency of anxiety*"): "*Once you start thinking, not only do you try to figure the best thing to do in the face of several possible next moves that a predator (including a social predator) is likely to make, you also think about what will happen if the plan fails.*"[13] So, on the one hand, imagining scenarios benefits the individual by providing options, offering room for better decisions and assessments in terms of survival and reproduction. If the tiger goes right, I might climb on that rock; left, I run to the tree. But, on the other hand, some of these scenarios display undesirable features, fuelling anxiety. What if the branch I am sitting on breaks, and I fall?

What If…? If Only I Had…

Let's take a look at what Daniel Kahneman and Dale T. Miller have termed "*counterfactual thinking*". Counterfactual thinking is the phenomenon of "what if" reasoning. In order to understand our world and our choices, we often imagine how things could have been different. The divorcee may ask, "*Would I have been happier today if only I had married someone else?*" A floundering student may ponder, "*If only I had chosen another major, maybe I would have a better shot at getting into graduate school.*"

In short, counterfactual thinking allows us to mentally simulate alternatives to our current reality in order to understand what features were most causal in

---
[13] LeDoux, J., 1998

bringing about the current outcome. A person may imagine how an outcome could have turned out differently, and can reflect on how the events that led to the current situation might have been different: "*If only I had…*"; "*I almost got it…*"

Let's look at an illustrative story by Kahneman and Tversky[14]:

> *Mr Crane and Mr Tees left town in the same limousine to catch different planes at the airport, leaving at the same time. Because of a traffic jam, they arrived at the airport 30 min after the official time of departure, and they both missed their flight. However, although Mr Tees's flight departed on time, Mr Crane's flight departure had been delayed by 25 min. This meant that Mr Crane arrived 5 min too late.*

When people were asked who was more upset, Mr. Crane or Mr. Tees, 96 % of the respondents declared that surely Mr. Crane was more upset, even though both men missed their flight and are now suffering the same inconvenience of having to reorganize their day.

Kahneman and Tversky suggested that the standard emotional script calls for both travellers to think of how close they were to making their flight. In this story, it is easier for Mr. Crane to imagine how he could have saved the crucial five minutes than for Mr. Tees to imagine how he could have gained the thirty minutes that separated him from his flight. If only he had not stopped to put up his umbrella! If only he had asked the limousine driver to take the other route to the airport!

When Bronze Feels Better Than Silver

A 1995 article about Olympic medallists provides fascinating insight into counterfactual thinking.[15] According to Victoria Husted Medvec, Northwestern University, Scott F. Madey, Cornell University, and Thomas Gilovich, Professor of Psychology at Cornell University, people's emotional responses to events are influenced by their thoughts about "what might have been". In particular, an analysis of the emotional reactions of bronze and silver medallists at the 1992 Summer Olympics, both at the conclusion of their events and on the medal stand, indicates that bronze medalists tend to be happier than silver medallists. The authors attribute these results to the fact that the most compelling counterfactual alternative for the silver medallist is winning the gold, whereas for the bronze medallist it is finishing without a medal. The silver medal could have been gold with a single step difference.

---
[14] Kahneman, D. and Tversky, A., 1974
[15] Medvec, V. H., Madey, S. F. and Gilovich, T., 1995

So sad :-(. By contrast, the bronze medallist feels happy for having made it onto the stand :-).

The Charismatic Spanish Bullfighter Yiyo

As reported by the psychologists Dale Miller and Brian Taylor,[16] the tragic and sudden death of the charismatic Spanish matador Yiyo provided all the ingredients necessary for a perfect counterfactual story. In 1985, Yiyo was killed by a bull just after replacing another matador near the end of the fight.

What if the charismatic Spanish matador Yiyo had not replaced the other matador?

The fact of Yiyo replacing the other matador unexpectedly, something that could just as easily not have happened, generated much more anguish and stress than the less improbable scenario of Yiyo dying at the horns of a bull which he had long been scheduled to fight.

In typical counterfactual thinking, events preceded by exceptional actions (replacing the matador at the last minute) are more easily imagined otherwise (what if…?) and generate more feeling than events following routine sequences.

Imagine the frustration generated by having correctly spotted four numbers out of five at the weekly lottery! While the holder of such a ticket hasn't really been any more unlucky than the millions of other losers, his emotional experience of not winning is very different and a lot more painful. So close and yet so far! If only he had stopped to think before choosing that one unlucky number!

Hindsight Is 20/20

This tendency to revisit the past in the light of subsequent events is common enough. We are so tempted to try to make sense of the past—such as trying to work out whether the attack on the Twin Towers could have been averted, or whether Chamberlain could have avoided the Second World War—that we may fail to recognize that this knowledge *now* will not have any effect on us, since we cannot undo what has already happened.

Because we know what did happen, we try to reorganise what we already know about the subject so as to let the outcome appear more or less inevitable or predictable than it really was at the time. Looked at in retrospect, the results of political elections seem more likely than they were at the time.

---

[16] Miller, D. and Taylor, B. R., 2002

We should learn from the past but be aware that, at that time, the context may have been uncertain and information sparse.[17] As stressed by Baruch Fischhoff, Professor in the departments of Social and Decision Sciences and of Engineering and Public Policy at Carnegie Mellon University: *"the lesson to be learned from the surprise attack on Pearl Harbour is that we must accept uncertainty and learn to live with it"*.[18]

In *The Black Swan*,[19] the bestselling author Nassim Nicholas Taleb quotes the example of a diary, written in Paris by the journalist William L. Shirer in 1940, as an illustration of the fact that it is easier to be wise and understand the world after the event. Indeed, Shirer was describing events as they took place rather than after. He was seeing events clearly as they happened, but did not know what would happen next.

Reading the diary with the benefit of hindsight, it should have been obvious that Adolf Hitler represented a terrible threat.

You'd Never Forgive Yourself

You mean all this time we could have been friends?

From *What Ever Happened to Baby Jane?*, a 1962 American psychological thriller film directed by Robert Aldrich, starring Bette Davis and Joan Crawford.

Bitter experience has shown that, if possible, one should strive to avoid making decisions that will only be the cause of regret later on. "What if" thinking has much to do with avoiding potential regret. Samuelson and Zeckhauser[20] consider the following situation for parents with a baby.

Most parents would never consider leaving their baby alone, sleeping in its crib, while they took a fifteen-minute car trip to run an errand. In the extremely unlikely case that the child was killed in a fire, the parents would feel tremendous regret and guilt; they would never forgive themselves. At the same time, most parents would not hesitate to take the child along in the car. Though the child is actually in much more danger in the car, the element of guilt associated with a car accident would be considerably less. Yet most of us, even when we understand the relative risk, will make the same call. The alternative seems unthinkable.

---

[17] Fischhoff, B., Watson, S. R. and Hope, C., 1984; Taleb, N. N., 2004
[18] Fischoff, B., 2003
[19] Taleb, N. N., 2010
[20] Samuelson, W. and Zeckhauser, R., 1988

Monkey See, Monkey Do

Now, when our mind is craving for clues, what better source of inspiration than what others do?

Social Proof

According to Robert B. Cialdini, we view any behavior as correct in a given situation depending on how much we see others engaging in it.[21] He referred to this as "*social proof*", a concept related to similar notions like "*group thinking*" and "*herd behavior*". It is assumed that adopting this behavioral strategy is more likely than not to lead one to do the right thing.

"*Everyone is doing it*" provides a clue in the form of a queue outside a nightclub as to how popular and trendy it is. After all, why would all those people want to go if it wasn't absolutely great? This sort of situation leads to a self-fulfilling prophecy. If everyone wants to go there, of course it is going to be more fun and exciting than the half-empty place down the road!

We move to a new neighbourhood and wonder what to do with our waste. The best way to find out is to see what the neighbours are doing. Should we let our towels be replaced every day in the hotel room? If the hotel places a discreet notice reminding us that environmentally-aware guests don't leave their towels to be washed every day, we will probably use our towels more than once too.[22]

Panhandlers and people collecting for charity on the street understand the impulse to fit in with what everyone else is doing when they "seed" their cup or collecting receptacle with a few coins to give potential donors the right idea. So do the producers of television sit-coms when they introduce canned laughter so as to influence our perception of the show.[23]

Milgram's Experiment: On 42nd Street, Look Up At the Sky and Everyone Will Follow Suit!

In the 1960s, the famous psychologist Stanley Milgram conducted a social contagion study on 42nd Street, New York City. Varying numbers of passers-by (all in fact colleagues of Milgram) stared at a sixth-floor window. As the number of passers-by increased, the percentage of people who stopped to stare also increased. The result was that 45% of passers-by stopped if one person

---

[21] Cialdini, R. B., 1984
[22] Thaler, R. and Sunstein, C. R., 2008
[23] Cialdini, R. B., 1984

was looking up, whereas 85% stopped if 15 were looking up. In other words, the more people were staring at the window, the more likely it seemed that there was in fact something noteworthy to look at.[24]

You can find amazing videos on the Internet. In one of them, an individual starts a weird, lonely dance in a park. Sun-bathing or sitting on the lawn on this sunny day, people look on. You have to visualize the odd situation of this guy, totally alone, dancing in such a strange way, and stared at by a whole host of people lying on the grass. It takes some long minutes before, very slowly, a few people join the dance in front of everybody else. As time goes on, there comes a point where suddenly a kind of threshold is reached, and it appears that the whole population of the park is now joining the festivities. The scene is unbelievable: everybody doing that crazy dance for no other reason than social imitation.

Herd behavior describes how individuals in a group can act together without planned direction. Herd behavior can lead to dramatic situations. For example, imagine a fire alarm going off in a closed, crowded environment, such as a rock concert or a disco. The likelihood is that everyone will rush to the same exit at the same time in a dangerous stampede rather than trying to figure out where their own nearest exit might be. Unfortunately, there are many examples from all over the world of tragedies that have occurred in such circumstances, when the herd instinct simply takes over.

On a more trivial level, consider the rush to buy this year's must-have item for Valentine's day or Christmas day, or the fashionable and expensive branded jogging shoes that have just been released to the shops. The sight of hordes of frantic shoppers struggling to get their hands on the desired items makes other shoppers want to get in on the action too. Why would so many people want them if they weren't absolutely fantastic? In his book, *Influence: the psychology of persuasion*, Robert Cialdini provides an astonishing picture of a shop assistant filled with disgust after the passage of enraged customers struggling for shoes, with no regard for whether they are the right size![25]

Would "Do As Others Do" Stand At the Root of the Evolution of Cooperation?

What are the advantages of doing as others do? Cooperation between humans made possible their bewildering increase and spread around the globe. But how could cooperation emerge? We now expose the role of reciprocity as a building block for cooperation. To do this, we shall make a detour around

---

[24] Milgram, S., Bickman, L. and Berkowitz, L., 1969
[25] Cialdini, R. B., 1984

the famous mathematical game called the "*prisoner's dilemma*" to arrive at the reciprocal "*tit-for-tat*" strategy which lies at the heart of cooperation.

In the obituary reporting the death of mathematician Albert William Tucker, Princeton University communication service relates the origin of that game. In 1950, addressing an audience of psychologists at Stanford University where he was a visiting professor, Tucker created the *Prisoner's Dilemma* to illustrate the difficulty of analysing non-zero-sum games, that is, scenarios in which one contestant's victory is not necessarily the other contestant's defeat. We report the description from the Princeton University communication service website.[26]

> *The Prisoner's Dilemma depicts two partners in crime confronted with the following choices: if one confesses and the other does not, the confessor goes free and the other goes to jail for a long time; if neither confesses, each goes to jail for a short time; if both confess, each goes to jail for an intermediate length of time. Each reasons that he is better off confessing because if the other confesses, he receives an intermediate sentence by confessing and a long sentence by not confessing; if the other does not confess, he goes free by confessing and receives a short sentence by not confessing. Since each reasons this way, each confesses, and so each is given an intermediate sentence; whereas if each had not confessed, each would have received a short sentence.*[27]

 A Small Dose of Theory: The Evolution of Cooperation

In the so-called "prisoner's dilemma", cooperation is not the rational choice. However, cooperation can emerge when individuals have to interact during an indefinite amount of time. In 1981, the political scientist Robert Axelrod and the evolutionary biologist William D. Hamilton co-authored an amazing paper[28] entitled "*The evolution of cooperation*". They described a kind of tournament. Any strategy competing in that tournament can make two moves 'cooperate' or 'defect', depending on all previous moves within the game.

Building on simulations of the tournament and on mathematical arguments from repeated games theory, Axelrod and Hamilton stressed the success and the central role played by the so-called "tit-for-tat" strategy. Proposed by the psychologist Anatol Rapoport, "tit-for-tat" strategy cooperates systematically on the first move and, on the following moves, systematically does the same (reciprocates) as the other player did on the previous move.

---

[26] https://www.princeton.edu/pr/news/95/q1/0126tucker.html
[27] https://www.princeton.edu/pr/news/95/q1/0126tucker.html
[28] Axelrod, R. and Hamilton, W. D., 1981

If we inherited the features of the tit-for-tat strategy, this could explain the "*social proof*" bias, that is, the way we pick up signals from our fellows and echo them.

Benign herding behaviors often crop up in terms of everyday decision-making. Consider the man on a business trip to a new city who has to decide between two restaurants A and B. They both look nice but they are both empty, so he chooses one of them, restaurant A, at random. Shortly after he is served, a couple walks down the street in search of a restaurant. Seeing that one has a customer who seems to be enjoying his meal, while the other is completely empty, they also choose restaurant A. The trend continues throughout the evening, with the result that restaurant A has a successful day's business, and restaurant B does not!

I Am Not Superstitious, That Brings Bad Luck

When our mind has embarked upon "what if" scenarios, or is looking for social clues in the behavior of conspecifics and copying them, is it trying to regain some control over events?

Burrhus Frederic Skinner, the late American psychologist (and famous for his radical behaviorism), discovered in a series of experiments, that even pigeons can be superstitious.

> *A pigeon is brought to a stable state of hunger by reducing it to 75 % of its weight when well fed. It is put into an experimental cage for a few minutes each day. A food hopper attached to the cage may be swung into place so that the pigeon can eat from it. (…)*
>
> *If a clock is now arranged to present the food hopper at regular intervals with no reference whatsoever to the bird's behavior, operant conditioning usually takes place. One bird was conditioned to turn counter-clockwise about the cage, making two or three turns between reinforcements. Another repeatedly thrust its head into one of the upper corners of the cage.*
>
> *The experiment might be said to demonstrate a sort of superstition.*
>
> *The bird behaves as if there were a causal relation between its behavior and the presentation of food, although such a relation is lacking.*[29]

Observing a pigeon's behavior once it has been trained to expect some food to be delivered at intervals, whatever the attitude may be in-between *reinforcements* (by food delivery), Skinner made the point that the pigeon appeared to

---

[29] Skinner, B. F., 1948

adopt a kind of superstitious behavior or magical thinking. That is, it falsely related a particular behavior to the appearance of food.

During the 2014 soccer FIFA world cup, we observed some interesting attitudes from players which could likewise be attributed to magical thinking. For instance, soccer players systematically spit on the ground when they miss a shot at goal or a pass (as if they wanted to expel the bad luck or the mistake?). In a similar way, the "guilty" soccer players systematically raise both hands, palm forward, as if to clear themselves of responsibility for a fault against an opponent, even if the fault appears effective on replay (hoping to influence the referee?).

We're All Control Freaks

> *"You sound like a control freak."*
> *The words are out of my mouth before I can stop them.*
> *"Oh, I exercise control in all things, Miss Steele," he says without a trace of humor in his smile.*
> *I look at him, and he holds my gaze steadily, impassive. My heartbeat quickens, and my face flushes again.*
> E. L. James, Fifty Shades of Grey, 2011

The term *"illusion of control"*, used by psychologist Ellen Langer,[30] refers to people's innate tendency to overestimate the extent to which they can control events, even those that they clearly have no influence over. This belief affects their behavior in a wide range of fields, from gambling to beliefs in the supernatural, to decisions that we make about our lives and careers.

Langer explains the illusion of control as resulting from people basing their beliefs about control on what she referred to as *"skill cues"*, or features of any given situation that are typically associated with skills—competitiveness, familiarity, and individual choice.

Control certainly gives a sense of safety that makes us feel protected from unfavorable events. In *Nature's Mind*, Michael Gazzaniga stresses the importance and the benefits of *"perceived control"* in nearly every aspect of daily living. Exerting control over the environment always had and still does have an adaptive advantage, compared to "letting go".

Bad Things Don't Happen to People Like Us

It seems as if people see themselves as immune to hazards.

---
[30] Langer, E., 1975

"*It won't happen to me*" is what most of us feel, even when we understand that we have a strong genetic predisposition to cancer or diabetes, or when we knowingly take part in activities that are bad for our health and safety.

A review of the literature[31] by Frances K. Barg, University of Pennsylvania, and Sonya A. Grier, American University at Washington D. C., revealed that African American women with breast cancer underutilized breast mammography screenings, resulting in poor health and high mortality. Barg and Grier often heard, as a justification for not using screenings, that cancer "*won't happen to them*", making it quite a struggle to convince them.

On another register of optimism, most of us think that we are better than average drivers. 79 % of young drivers think they are better than the others, which is clearly impossible. 93 % of American drivers rate themselves as better than average.[32]

Similarly, people tend to think they are more likely than average to live longer than 80 years.[33]

Magical Thinking

The illusion of control is stronger in situations that are highly competitive, such as the world of finance. People tend to see themselves as responsible for events, even when there is no link between their behavior and what has happened. It is easy to think of examples of this illusion in a casino: people kiss the dice "for luck" or throw them harder when they want to get a higher score; the would-be Lothario who got lucky the last time he went to a nightclub remembers the stylish purple shirt he wore that evening, and decides to wear it every time he is in the mood for female company; the checkers enthusiast has a meal of fried fish before winning a series of games, and decides always to eat fish before a checkers tournament from now on.

Magical thinking is a quality of the way we live that continues to linger.

The tennis player Rafael Nadal always positions his bottles on the court with incredible precision. Many sports people display similar types of magical thinking.

James Frazer published *The Golden Bough* at the end of the nineteenth century. It is a wide-ranging volume that collects and compares magical and religious beliefs and behaviors from a variety of cultures and settings. Frazer provides descriptions of magical thinking, that is, the commonly held idea that people can influence the world around them by behaving in certain ways,

---

[31] Barg, F. K. and Grier, S. A., 2008
[32] Svenson, O., 1981
[33] Weinstein, N. D., 1980

and the tendency to see relationships between objects on the basis of coincidental physical similarities or through contact.

Until recently, and especially until the discoveries of Louis Pasteur, people were exposed from birth to many diseases that they could not cure. Diseases were passed on to other people by contact, so it seems quite understandable that we would develop a "magical aversion" to any contact with foreign objects, bodies, or fluids.

Homeopathy: The Magic of Almost Nothing

Homeopathy is a form of "alternative medicine" in which practitioners treat patients using highly dilute preparations. As such, homeopathy stands for magical dilution.

British best-selling science writer Ben Goldacre,[34] in his astonishing book *Bad Science*, explains how difficult it is to convince patients with the help of rational arguments, such as the huge amount of scientific findings which demonstrate that homeopathy is no better than a placebo. Homeopathy attributes an impact to one active particle diluted with 1,000,000,000,000 water particles. For instance, what is known as a 6 CH dilution ends up with the active molecule diluted by a factor of $100^{-6} = 10^{-12}$ (one part in one trillion or 1/1,000,000,000,000). Higher dilutions follow the same pattern.

In homeopathy, a solution that is more dilute is described as having a higher potency, and more dilute substances are considered by homeopaths to be stronger and deeper-acting remedies than less dilute. Goldacre notes that the typical dilutions that can be found in drugstores range from CH6 to CH40, and shows that a CH10 product already represents a higher dilution than one would find "*if there was a drop of the active substance in all the water in Lake Geneva*".

"Magic", or what anthropologists have termed "magical thinking", refers to the universal trend to seek symbolic and meaningful relationships between objects and events.[35]

Does God Answer Your Prayers?

Research by Dennis L. Jennings, Mark R. Lepper, and Lee Ross, all three at Stanford University, describes a striking example of how people make unproven associations:[36]

---

[34] Goldacre, B., 2009
[35] Frazer, J., 1993 (1890)
[36] Jennings, D. L., Lepper, D. L. and Ross, L., 1981

"Does God answer prayers?"
"Yes" says the layman..., "because on a lot of occasions I asked God for something and he gave it to me."
"Well," asks his skeptical friend, (...), "how many times did you ask and not get what you wanted?"

The conversation above discusses two possible outcomes, but there's also a third; no prayer and no favourable outcome. None of the three solves the issue of whether or not God answers prayers. A fourth possible scenario, no prayer and outcome, would need to be studied too, but most people don't even want to go there.

So, does God answer your prayers or did you just forget when he did not?

	God answered	God did not answer
Did pray		
Did not pray		

Picture: does god answer prayers? an objective view

Objectively, there are four possibilities depending on whether you pray or not, and whether God appears to answer or not. Those four possibilities are illustrated above by four rectangles.

Picture: does god answer prayers? a subjective view

Now, in the subjective assessment of the situation, the upper right (red) rectangle (God did not answer my prayers) is much smaller than the upper left (deep blue) rectangle (God answered my prayers). The other two (light blue) rectangles at the bottom are neglected in the appraisal: the cases where I did not pray and either God did or did not answer.

Was the German Bombing of South London Done At Random?

During the Germans' intensive bombing of South London in the Second World War, a few areas were hit many times over while others were not hit at all. The places that were not hit seemed to have been deliberately spared, and people concluded that those were where the Germans had their spies.

When the famous mathematician William Feller[37] analysed the statistics of the hits, he found that the distribution was random. For the layman's understanding, however, the bombing pattern could not be accepted as random. It didn't fit with the story.

The World Did Not End In 2012, But the Mexican Tourism Industry Flourished!

During 2012 there was a lot of excitement about the impending "end of the world", supposed to have been predicted by the ancient Mayans for December that year. Those who believed that the world would end, or that there would be some major world change, saw signs and symptoms in everything that happened that seemed to link with the narrative behind the story, and confirm that it was likely to be true. Perhaps the Mexican tourism industry was one of the major beneficiaries of the "theory"!

 A Small Dose of Conspiracy Theory

On the one hand, the ability to identify relevant patterns offers an adaptive advantage thanks to better predictions. Our ancestors who spotted regularities or signs around them—clues to the a possibility of a storm or the coming of the seasons, a sudden silence in the vicinity indicating the presence of a large predator, body and behavioral signals when a woman is in her ovulating phase, traits revealing the quality of seeds, characteristic features of a safe place to stay—had an advantage in terms of survival and reproduction.

---

[37] Feller, W., 1968

On the other hand, as we have discussed in "the paranoid optimist theory",[38] seeing intentions behind events also offers an adaptive advantage. The costs of imagining an agent behind random occurrences are less than those of ignoring a potential agent. Indeed, as CSI Grissom points out (see the beginning of the chapter), the cavemen who believed that a movement in the savannah grass was due to the wind carried genes that left fewer heirs to testify.

So our magical thinking may be a consequence of our adaptively relevant propensities to look for agents and patterns, at the expense of truth. Pascal Boyer, anthropologist and Professor at Washington University in St. Louis, states that "religious concepts are parasitic upon other mental capacities".[39]

---

[38] Haselton, M. G. and Nettle, D., 2006
[39] Boyer, P., 2001

6
Images Call More to Mind Than Words and Numbers

As for the terrified Spanish soldiers entering into unknown territories, our guts sometimes take over. In the following stories, you might encounter your deeply rooted emotions, hunger, sex, and danger, fear of air travel, and bloodstained news. Walking on sunshine and reading terrifying anecdotes, you will almost bite a worm in an apple and chill witnessing Jack's skull being crushed by a semi-trailer… Finally, after a long series of reds, you will be craving for the black to come out at roulette.

> *As for the Spaniards, Atabalipa (Atahualpa) heard from Indians who were spies, that the Spaniards had all gathered in a shed, full of fear, and that none of them had shown himself on the open place; and it was the truth, what was said by the Indian, because I heard many Spanish were urinating on themselves from pure terror without feeling it.[1]*

Picture: painting of Atahualpa with the "conquistadores"

---

[1] Pedro Pizarro. The original Spanish text reads: *Pues estando así los españoles, fue la noticia a Atabalipa, de indios que tenía espiando, que los españoles estaban todos metidos en un galpón, llenos de miedo, y que ninguno parescía por la plaza; y á la verdad el indio la decía, porque yo oí a muchos españoles que sin sentirlo se orinaban de puro temor.*

Thinking With Our Guts

On November 16, 1532, In Cajamarca, the *Conquistadores* Faced Unknown Fear

Humans fear known risks, and they also fear the unfamiliar, simply because they do not know how they should react to it. We have a vivid historical description of such fear when, on November 16, 1532, in Cajamarca, in modern Peru, the Spanish conquistador Francisco Pizarro was about to encounter the Inca emperor Atahualpa. Some of the conquistadors admitted to the terror of facing the unknown.

Like the terrified Spanish soldiers during the Spanish conquest of what we now know as Latin America, emotion can overwhelm the rational control of our selves. This illustrates what Loewenstein and Thaler[2] call "*visceral thinking*", namely, immediate visceral reactions such as worry, fear, anxiety, and dread, when faced with dangerous situations.

Picture: the quiet city of Cajamarca nowadays, in modern Peru

So far as many people from the West are concerned, Latin American countries remain associated with risks and danger. In order to alleviate or even reverse such views, Colombia even played on the notion of risk in one recent advertisement:

Colombia, the only risk is wanting to stay.

---
[2] Loewenstein, G. and Thaler, R., 1989

The Fast Track to Fear

Significantly, there is a constellation of casual terms related to fear. Joseph Ledoux lists several of them:

alarm, scare, worry, concern, misgiving, qualm, disquiet, uneasiness, wariness, nervousness, edginess, jitteriness, apprehension, anxiety, trepidation, fright, dread, anguish, panic, terror, horror, consternation, distress, unnerved, distraught, threatened, defensive.[3]

In modern Chinese, fear is written as a combination of two symbols, each one expressing a different level of fear: "*panic*" on the left as well as "*common fears*" on the right.

Picture: fear in modern Chinese

So, what is fear?

 A Small Dose of Theory on Fear

> The system that detects danger is the fundamental mechanism of fear, and the behavioral, physiological, and conscious manifestations are the surface responses it orchestrates.[4]

> Joseph LeDoux adds that in reptiles, birds, and mammals, the brain performs the function of defense against danger "*using a common architectural plan*". Such similarity is the symptom of a deeply rooted and crucial priority for living organisms: protecting oneself from danger.

Due to the "top of the list" priority of detecting danger, brains have been shaped for that purpose. In particular, we are wired to assess whether "something" is

---
[3] Ledoux, J., 1998
[4] Ledoux, J., 1998

good/harmless or bad/dangerous, before knowing what that "something" is. Emotions, moods, and feelings appear as quick hard-wired assessments of a situation. They are complementary to slower, more intellectual, appraisals. These two evaluation pathways correspond to pathways in our brain.

 A Small Dose of Theory On Fear Neural Circuits

When a stimulus is processed by our brain, is takes two pathways, the quick and dirty "what like road" and the slower "what road", so that we know "what something is like", good or bad, before we know "what something is", a stick or a snake. The fast pathway goes to the amygdala, a deeply located, sub-cortical, almond-shaped zone that handles fear responses. By this route, our mind quickly picks up a crude image of the external world. Another road goes to the cortex, which yields a more detailed and accurate image, but at the expense of taking longer. The adaptive advantages of this quick thalamic pathway are overwhelming for survival.

Joseph LeDoux stresses that the time saved by the amygdala in acting on the thalamic information, rather than waiting for the cortical input, may be the difference between life and death. It is better to have treated a stick as a snake than not to have responded to a possible snake.[5]

Picture: better take a stick for a snake than a snake for a stick

---
[5] Ledoux, J., 1998

The Slow Road to Thinking, Reasoning, and Consciousness

So we can have an emotional answer, the "*what like*", before knowing exactly "*what*" we are confronted with, because emotions bypass the neocortex to get to the amygdala.[6]

Therefore, now that we understand somewhat better the structure of our brain wiring, we also understand better why we sometimes feel torn between a visceral immediate feeling, "*I feel that*", and a slower, more elaborate assessment, "*I think that*".

But does this mean that we can "tame" our fear system? Can we control our emotions? Do we have to? In fact, it seems hard to dominate our emotions because of the nature of the connections between the cortex and the amygdala.[7] Indeed, connections are far weaker from the cortex to the amygdala than the opposite. As LeDoux puts it, "*This may explain why it is so easy for emotional information to invade our conscious thoughts, but so hard for us to gain conscious control over our emotions.*"

Emotions and feelings have high adaptive value. Moreover, and contrary to widespread opinion, they are a substantial and crucial component of sound decision-making.[8]

Fear of Flying

Planes and cars are recent developments in human history, and people are much more likely to die in cars than in planes. So why is it that we are so afraid of planes? This is all about technology, so surely it should be easy to be rational? But flying in airplanes triggers ancestral fears, regardless of reassuring statistics.

Imagine that you are driving a car, and thinking about the possibility of having an accident. Generally, your impression is that it is highly probable that, in a car accident, you would escape unscathed.

What's going on?

While people may have access to the statistics on car safety (and sometimes choose not to pay heed to them), they rely more on emotional responses to the situation. A car "feels" safer than it really is because it travels on the ground, and therefore seems more ordinary, more in line with what is familiar to us.

The driver and passengers in a car do not perceive themselves to be in a dangerous situation. They can see the familiar world going past outside just as it would if they were walking, albeit more quickly. Whether accurately or

---

[6] Ledoux, J., 1998
[7] Ledoux, J., 1998
[8] Bechara, A., Damasio, H. and Damasio, A. R., 2000

not, their perception is that, if something bad happens, they will be able to get away from it because they are in an environment that they know about, one that provides parameters within which they know how to operate.

Picture: flying above the Andes

On the other hand, people tend to overestimate the risk associated with air travel, even though statistics show that travelling by car is actually much more dangerous per unit distance travelled. According to statistics presented by Wikipedia[9], the risks (measured in deaths per billion kilometres) associated with a particular long-range journey from one city to another are more than sixty times higher by car than by air.

However, the airplane passenger is generally of the impression that, in the event of something bad happening, there is no way to escape. When he looks out the window (if he is not stuck in the middle of a long row) he does not see the familiar sights of horizon and landscape; he sees clouds, or perhaps the land or a glittering ocean, far below.

Human beings have been travelling by air for only a minuscule fraction of the time they have existed on Earth. There is no instinctive frame of reference for the person in this situation. This feeling of apprehensiveness, of being cornered, does not subside even though he knows, having read all about it in well-respected periodicals, that, statistically, many more people die in car accidents than in airplane crashes. Still, the feeling of helplessness triggers fear. In a plane, the typical reaction patterns of flight, fight, or play dead are no longer useful or possible strategies. If something bad happens, there will be *nowhere to go*. It is as if the hunter-gatherer pursued by a tiger finds himself in a blind alley with no escape, and no option but the unpromising "play dead". There are no rocks to hide behind, no trees to climb. Seeing no apparent escape route, our nervous air traveller feels cornered, apprehensive, and claustrophobic.

---

[9] http://en.wikipedia.org/wiki/Aviation_safety#cite_note-33

The cult movie *Snakes in a plane* combines in a terrifying cocktail the hardwired fears of snakes and of being trapped. The theatrical release poster reads:

Sit Back. Relax. Enjoy the fright.

What Is the Riskiest Part of a Plane Trip? Driving to the Airport!

Compare the anxiety that many people fear when flying in a commercial airline with the tranquillity the same people usually feel when driving a car: just as one often sees nervous travellers shovelling tranquilisers into their mouths before boarding a plane, one often observes that most people will jump into their car and join the traffic on a busy road without giving it a second thought.

In his study on the perception and management of risk, Andrew Stewart[10] reported statistics issued by Barry Glassner[11] according to which fewer than 13,000 people have been killed in the entire history of civil aviation accidents since 1914 in the US, compared to three times that number in car accidents in the US *every single year*![12]

This simple comparison of facts and beliefs shows us that our opinion about how risky any given situation is may well be governed by emotional factors, despite the statistics. Cars are a lot more dangerous than planes. However, as Stewart puts it, *"What we feel often wins out over any objective analysis that we are presented with."*[13]

Feelings change our perception of the riskiness of a given situation. Because, in the case of flying, feeling dominates over cognitive evaluation, there is no point in reassuring the nervous flyer that the chances of being killed in a plane crash are one in millions. He will not be reassured by statistics or probabilities because his instinct for risk has taken over, and statistics have lost any meaning for him.

As a matter of fact, people are willing to pay more for airline travel insurance covering death from terrorist attack than from all possible causes despite the fact that, unlikely as it is that they will die in a plane crash, it is even less likely that the crash will result from terrorist activity.[14]

---

[10] Stewart, A., 2004
[11] Glassner, B. *The Culture of Fear: Why Americans are Afraid of the Wrong Things*. New York: Basic Books, 1999
[12] Stewart, A., 2004
[13] Stewart, A., 2004
[14] Sandman, P. M., Miller, P. M., Johnson, B. B. and Weinstein, N. D., 1993

Insurance Feelings

Although insurance ought to be a dispassionate business, a matter of hard numbers and mathematics, emotions play an important role in the decisions people make about what to insure and how much things are worth.

In a study by Hsee and Kunreuther,[15] it was found that people were willing to pay twice as much to insure a beloved antique clock (that no longer works and cannot be repaired) against loss in shipment to a new city than to insure a similar clock for which *"one does not have any special feeling."* (In the event of loss, the insurance paid $ 100 in both cases.)

According to Howard Kunreuther, from University of Pennsylvania, when they were asked about their insurance decisions and investment in protective measures, subjects in both laboratory and survey studies were not generally inclined to worry about risks that were unlikely to happen[16]:

> *Even after Hurricane Andrew, most residents in hurricane-prone areas along the Atlantic and the Gulf Coasts appear not to have been investing in loss reduction measures.*

This is understandable; after all, people have a limited reservoir of attention and energy to expend and there simply isn't time to worry about everything that might happen. What people *believe* is likely to happen is heavily influenced by what they are told and what they see, in contrast to what is *actually* likely to happen. And these beliefs are generally formed by emotional responses and not by a rational reaction to the cold, hard facts of statistics. When asked about their insurance decisions, subjects in both laboratory and survey studies indicated a disinclination to worry about low-probability hazards,[17] a strategy that is easy to understand when one considers the fact that, if people were to worry about low-probability but high impact hazards (such as hurricanes), a huge proportion of their time, energy, and attention capacities would be taken up. Quite simply, unless we ignore many low-probability threats, daily life would become impossible.

People tend to think of the insurance that they pay for as an investment. Making claims and receiving payments (by insuring against more probable losses) seems to be viewed as a return on the premium. Insuring against hazards that "don't occur" because they are so infrequent that individuals have generally no experience of them seems to be a waste of money, so such things are usually left uninsured.

---

[15] Slovic, P., Finucane, M. L., Peters E., and MacGregor, D. G., 2002
[16] Kunreuther, H., Mitigating Disaster Losses through Insurance, Journal of Risk and Uncertainty, 12:171–187, 1996
[17] Kunreuther, H. and Slovic, P., 1978

"When It Bleeds It Leads!"

The journalists' phrase, "*When it bleeds it leads*", refers to the fact that newspapers love to cover stories about disasters. Here again, dramatic deaths sweep dry figures away.

Barbara Combs and Paul Slovic[18] studied two newspapers from opposite coasts of the United States and found that they had similar biases in reporting life-threatening events. Statistically frequent causes of death—and therefore intrinsically more risky to the average reader—such as diabetes and cancer, were rarely reported, while violent, catastrophic events such as tornadoes, fires, homicides, drowning, car accidents, and so forth were reported much more often than less dramatic events that resulted in a similar number of deaths. People are about sixteen times more likely to die of disease than in a car crash, but car crashes were reported three times more frequently.

When Blackheads Matter More Than Cancer

In 1989, George Loewenstein and Richard Thaler[19] published a study showing anomalies in the apparently rational behavior of human beings, indicating that our behavior actually seems to reveal extremely short-sighted preferences. No matter how intelligent people may be, they find it difficult to consistently make the wisest decisions about what to do. We can see this sort of short-sightedness when a dermatologist complains about how her patients react to her warnings of the risks to skin health:

> *My patients are much more compliant about avoiding the sun when I tell them that it can cause large pores and blackheads than when I say it can cause cancer!*

We Do Not See a House But a Handsome Or an Ugly house

Visceral reactions have been studied by scholars of social psychology for years, especially since the pioneering work of late social psychologist Robert Zajonc, and have begun to attract renewed attention in recent years. According to Zajonc, all perceptions contain some feeling. "*We do not just see* "a house": *We see a handsome house, an ugly house, or a pretentious house.*"[20]

---

[18] Reported by Slovic, Fischhoff and Lichtenstein p. 468 of Kahneman, D., Slovic, P. and Tversky, A., 1982
[19] Loewenstein, G. and Thaler, R., 1989
[20] Zajonc, R. B., 1980; Zajonc, R. B., Adelmann, P. K., Murphy, S. T. and Niedenthal, P. M., 1987

Picture: a typical handsome "*maison de maître*" near Bordeaux in the South West of France, in 2005, along with the dog "Myrtille"

We quickly and automatically label what we perceive as good or bad, bringing us back to the wiring of our brain, with the slow and accurate pathway to the cortex versus the quick and dirty one to the amygdala. Indeed, Joseph LeDoux writes that[21] "*The emotional meaning of a stimulus can begin to be appraised by the brain before the perceptual systems have fully expressed the stimulus. It is, indeed, possible for your brain to know that something is good or bad before it knows what it is.*"

In the same academic paper, Robert Zajonc and his Stanford University team explored how married couples, who had been together for 25 years, had begun to develop similar facial features. The study involved 110 participants (55 couples) whose photographs were taken in their first year of marriage and 25 years after. It was observed that long standing couples started to resemble each other more than they did when they got together.

A possible explanation is that a lot of human emotions and feelings are expressed through the face, and that when two people make similar facial expressions for 25 years, it can result in similar wrinkle patterns. Most married couples who have been together for 25 years or longer can identify with the other person's feelings through empathy. Same life… same feelings… same wrinkles… same faces.

---

[21] Ledoux, J., 1998

Chickening Out

In their paper *Risk as Feelings*,[22] George F. Loewenstein, Christopher K. Hsee, Elke U. Weber and Ned Welch recall that even people who are not suffering from full-blown phobias commonly experience powerful fears about outcomes that they recognize as highly unlikely (such as airplane crashes) or not objectively terrible (such as public speaking); in contrast, many experience little fear about hazards that are both more likely and probably more severe (such as car accidents).

Our feelings influence almost every decision we make. In one simple experiment carried out by psychologists David Dunning at Cornell University, Leaf Van Bowen from the University of Colorado Bolder, and George Loewenstein[23], students were offered $ 1 in exchange for telling a joke in front of a class the following week. When the appointed time arrived, all the students were given the opportunity to change their minds. As their fear increased with the proximity of their rendition, many students "chickened out": 67 % of those who initially volunteered to tell a joke (6 out of 9) decided not to when the time came, while none of those who had initially declined the offer (0 out of 49) changed their mind.

Our Mental States Blow Hot and Cold

When aroused by hunger, sex, or anger, people fail to predict how they will behave in a "cool" state, and when they have calmed down they fail to predict the influence of arousal. In both situations, they underestimate the impact of a change from their current state.[24]

We are not good at estimating how we will feel and behave when the situation changes. The person who has been involved in a minor car accident and is full of rage towards the other driver will tend to assume that he is still going to be furious by the time he gets home. Instead, when he has had time to calm down and assess the damage, he is likely to feel a lot more forgiving.

Even more commonly, suffused with feelings of love in a post-orgasmic state, way too many of us assume that we're always going to feel as deeply in love, only to realise when our heart stops racing that our companion still snores, and still annoys us with his political views.

---

[22] Loewenstein, G. F., Weber, E. U., Hsee, C. K. and Welch, N., 2001
[23] Dunning, D., Van Boven, L., and Loewenstein, G. F., 2001
[24] Kahneman, D. and Thaler, R., 2006

Walkin' On Sunshine

Picture: five am, sunrise in Pinamar, Argentina; it is going to be a nice summer day!

Many experiments have found that, when they are in a good mood, people tend to make optimistic judgments, while those in bad moods make pessimistic judgments and choices. People who read sad newspaper articles subsequently gave higher risk estimates for a variety of potential causes of death (e.g., floods, disease) than people who read happy newspaper articles.[25]

In a research project by Eric J. Johnson, Professor of Business at Columbia Business School, and Amos Tversky, subjects were asked to read newspaper reports that described either accidents (fatal and non-fatal) or positive events, and were subsequently asked to estimate the chances of certain specific accidents happening to them and/or to the population in general. The subjects who had read "sad" reports were more likely to estimate risk as high, whereas those who had read "happy" reports were more likely to estimate it as low, even when the newspaper report was about an incident very different to the one they were asked to consider.

---

[25] Johnson, E. J. and Tversky, A., 1983

Customers Leave Larger Tips On Sunny Days

In 1993, Edward M. Saunders Jr established that days with higher cloud cover in New York are associated with lower aggregate U.S. stock returns.[26] Three years later, Bruce Rind[27] published a paper about two studies conducted at a casino hotel in Atlantic City that investigated beliefs about sunshine and tipping. People who believed that the weather was going to be good were more likely to leave bigger tips—in fact, it was demonstrated that they could be manipulated into leaving a bigger tip by their waiter writing a note on the back of their bill to the effect that sunshine was on the way!

Soccer Results Impact Wall Street

Stefano Dellavigna, Professor of Economics at University of California, Berkeley, assumes that, if mood can have a tangible result on behavior, mood-altering events should create measurable results. In fact, Dellavigna[28] reports on a 2007 investigation by Alex Edmans, Diego Garcia, and Oyvind Norli, in which international soccer matches impact the daily stock returns for the losing country.

The Peak-End Rule: Peak and End Experiences Matter More Than Duration

Retrospective evaluations of pain have been reported to follow the peak/end rule;[29] that is, they could be predicted by a simple measure of a peak pain level and a measure of the pain level at the end of the painful episode. However, the correlation between the duration of the pain and the retrospective perception of the pain was found to be relatively negligible.

In a study of patients undergoing colonoscopies, Donald A. Redelmeier, Professor of Medicine at the University of Toronto, and Daniel Kahneman explored the biases of memory in terms of recalling pain. Patients were inclined to make decisions about future treatments on the basis of how they experienced previous interventions. Redelmeier and Kahneman found that these memories were heavily influenced by the intensity of pain during the worst part of the treatment and towards the end of the experience. Patients' memories of the degree of pain they experienced tended to correlate with the

---

[26] Saunders, E. M., 1993
[27] Rind, B., 1996
[28] Dellavigna, S., 2009
[29] Frederickson, B. L. and Kahneman, D., 1993

levels of pain they had felt during the final 3 min of the procedure, and not with the level of pain experienced throughout the colonoscopy.[30]

Could something similar be operative in the poignant and painful experience related by Dan Ariely[31], Professor of Psychology and of Behavioral Economics at Duke University, in his best-selling book "Predictably Irrational"? Ariely suffered severe injuries from an explosion earlier in his life. He had to be treated daily with baths. Nurses and physicians were in favor of finishing the bath as fast as possible. They were generally opposed to deliver treatments by starting from the most painful part and then going to the less painful, which would have been better.

Amy M. Do, Alexander V. Rupert, and George Wolford, all three from Dartmouth College, Hanover, New Hampshire, report[32] on how a similar effect can be seen in the area of positive experiences. They cite work that showed research subjects rating wonderful lives that ended suddenly as better than "so-so" lives that lasted for much longer, the so called *"James Dean effect"*. For pleasure, peak and end matter more than duration.

On television, advertisements that create positive feelings are appreciated more by viewers when they have strong positive endings and very intense peaks. When people look back over their experiences, they tend to rely on their recollection of just a few distinctive moments. The peak/end rule can also be found in movies, popular songs, and books, all of which often start and finish in a dramatic way. Movies like the popular *James Bond* series typically begin and end with a bang.

"I Know a Brazilian Man Who...", Or the Power of Anecdotes and Vivid Testimonies

People seem to be much more sensitive to anecdotal, vivid evidence than to dry statistical facts. Where stories start by *"Once upon a time..."*, or when anecdotes begin with something like *"My close friend Henry told me..."*, they can convince us more than any figures, and "prove" a general proposition, despite their anecdotal uniqueness.

Richard Nisbett, Professor of Social Psychology at the University of Michigan in Ann Arbor, Eugene Borgida, Professor of Psychology and Law at the University of Minnesota, Rick Crandall, and Harvey Reed relate the following story which, in one form or another, will certainly have happened to all of us.[33]

---

[30] Redelmeier, D. A. and Kahneman, D., 1996
[31] Painful lessons, 30/01/2008; and also in Ariely, D., 2008
[32] Do, A. M., Rupert A. V. and Wolford, G., 2008
[33] Nisbett, R., Borgida, E., Crandall, R. and Reed, H., 1976

Imagine that you want to buy a new car, and decide to get a Volvo, the famously robust and reliable Swedish model. As a prudent, sensible buyer you inform yourself by reading authorised statistics and specialist magazines. New cars are expensive, so you want to make sure that you are getting good value for your hard-earned money! Then, the day before buying the car, you are at a cocktail party and a friend says that he has had to repair his Volvo's fuel injection system twice since buying it recently. What do you think is the most convincing: hard figures or your friend's personal experience? No bland figures can rival your best friend's advice on which car model to buy.

In terms of probability, this testimony adds one instance to the number N of Volvo owners, and shifts the average repair score up by no more than a jot. Statistically, it is practically irrelevant. You would need to know about the experience of thousands of Volvo owners in order to make an informed decision. Nevertheless, because of your friend's emotional, concrete, and vivid information, presented to you in compelling narrative form, you decide not to buy a Volvo after all.

When Two First Ladies Prompt More Cancer Detection Than Dry Statistics

In 1974, the waiting lists of cancer detection clinics in the US were exceptionally long, not because medical researchers had issued new, convincing statistics that showed the importance of detecting the disease early, but because Mrs. Ford and Mrs. Rockefeller, two first ladies, had just had such tests carried out.[34]

The "story" of the two influential women having themselves assessed for the disease was, to the American public, much more compelling than some dry statistical information about how many Americans were likely to develop cancer over the course of their lives. Because they knew about the two famous women—and probably felt as though they actually knew them, thanks to the widespread coverage of their activities in the media—the idea of having themselves screened for cancer seemed more sensible, more realistic, than if they had not heard the story.

NGO advertisements encouraging us to help the victims of famine emphasize less the numerical abstract of the victims as a whole than the moving faces of a few of them.

---

[34] Nisbett, R., Borgida, E., Crandall, R. and Reed, H., 1976

The Availability Heuristic

The "*availability heuristic*"[35] refers to the ways in which people predict the frequency of an event depending on how easily an example can be brought to mind. For example, a man might argue that smoking cigarettes is not unhealthy because "*my neighbour smoked three packs a day and he lived to be a hundred!*" or a young mother might argue that she does not believe that breast-feeding is the best choice for her, as "*I was bottle-fed by my mother and I am just fine!*".

As a result of his personal experience, the person in the first example believes that his neighbour's case is common, and maybe is even the norm, despite what he has read in the newspapers about the dangers of smoking, and despite any statistics. Because he knew his neighbour personally, and does not know the many people dying of smoking-related cancers, the former information seems to him to be much more relevant.

In the second example, the young mother extrapolates from her views of her own health to conclude that the overall statistically proven benefits of breast-feeding are not relevant to her case.

However, the neighbour's case may well be an extraordinary exception that has little or nothing to do with the general situation, as may the case of the new mother.

Another example of availability bias can be found in the persistent rumor that people who drive red cars get more speeding tickets. The belief stands firm even though insurance companies do not ask for the color of the vehicle when rating coverage. Several fanciful explanations have been given to explain this bias. It has been suggested that, because red is such an eye-catching and vivid color, red cars would tend to grab police attention. It has also been said that looking at the red color would accelerate a driver's pulse. But it may also be that, because many sports cars are red, like the famous Ferrari, red cars tend to go faster and tend to get speeding tickets more often.

Phil Healey and Rick Glanvill[36] even reported an urban legend in which police officers were systematically fining red cars and deliberately ignoring other colors, in the same way that snooker players would have looked for a series of red balls before striking other colors.

---

[35] Kahneman, D. and Tversky, A., 1974
[36] Healey, P. and Glanvill, R., 1996

Which City Is the Biggest? San Diego Or San Antonio?

When asked "*Which city has a larger population: San Diego or San Antonio?*" about two thirds of Americans responded correctly that San Diego is larger.[37] When a German group was asked the same question, a full 100% of the respondents answered correctly!

Why?

Although the Germans knew less about American cities in general, they inferred that the city whose name they recognized more readily must be the bigger one: they used a rule of thumb to make their analysis. In this case, as in so many, the rule of thumb turned out to be correct, as pointed out by Gerd Gigerenzer.

Why are we so ready to believe the evidence of a small number of people or factors with which we are familiar rather than a more objective analysis? The fact is that we are hard-wired to rely, as a rule of thumb, on the idea that what is familiar is most likely to be relevant. And this hasty analysis often makes perfect sense. As we saw, humans might have lived in groups of no more than about one hundred and fifty people, where familiarity was meaningful; it seems that statistics on a larger scale have less impact on our minds.

The French Eat Snails, Not Slugs

Gut feelings and disgust have much in common. You may know that French enjoy eating snails (not raw of course, but with garlic and butter, and after a lengthy preparation!). But why don't they eat slugs? Could it be that slugs carry diseases? Some scholars, such as Paul Rozin[38], Professor of Psychology at the University of Pennsylvania, argue that disgust might be an adaptation to avoid harmful food.

Picture: would you eat it?

---

[37] Goldstein, D. G. and Gigerenzer, G., 2002
[38] Rozin, P., Food for thought. Paul Rozin's Research and Teaching at Penn. Penn Arts and Sciences, Fall 1997, http://www.sas.upenn.edu/sasalum/newsltr/fall97/rozin.html

Picture: would you eat this Burgundy snail as French people do?

Paul Rozin invites us to consider this joke:

Question:

What's worse than biting into an apple and finding a worm?

Answer:

Biting into an apple and finding half a worm.

"The offensiveness of eating worms is one the most powerful things you can imagine," says Rozin.

People also tend to be reluctant to eat "joke" chocolates shaped like turds, even though they know very well that the item is made of nothing but delicious chocolate. This raises several questions: What does it mean when we say that something is disgusting? And why are some things disgusting while others are not?[39]

Most of us would be disgusted if we saw a fly or even someone's hair in our soup, as if the out-of-place item rendered the soup improper for consumption. However, our reaction to the foreign object may far outweigh the actual risk of contamination.

---

[39] Rozin, P. and Fallon, A., 1987

A Small Dose of Theory On Disgust

Paul Rozin and April Fallon, Professor of Psychology at Fielding Graduate University, approach disgust as a food-related emotion that they define as "*revulsion at the prospect of oral incorporation of offensive objects*". They point out that objects of disgust are virtually all of animal origin.

Humans are omnivorous and the benefits are clear-cut: when you get short of your traditional diet, you can turn to more "exotic" ones. But, as economists say, "there is no free lunch", and benefits come at a cost. The other side of the coin is known as the "curse of the omnivore". Indeed, when you stop exploiting your usual, tried and tested food stock, you have to look around. But how do you know that what enters your mouth is safe now?

As Jonathan Haidt, Professor of Ethical Leadership at New York University, Clark Richard McCauley, Professor of Sciences and Mathematics at Bryn Mawr College, and Paul Rozin[37] put it, "*oral disgust makes sense as an evolutionary adaptation for an omnivorous species living with the constant threat of microbial contamination.*" There are strong adaptive reasons for being cautious about what we ingest. The feeling of disgust is our safeguard, preventing us from potentially harmful food. That may be the reason why we are disgusted by those bugs and creeping animals which may have been in contact with spoiled food (e.g., cockroaches, rats, flies, worms).

Beliefs relating to tiny elements like the hair in the soup, or the finger nail in a dessert, are not unlike the beliefs that underlie the practice of homeopathy.

One of the most frequently used and best validated questionnaires of disgust assessment is the Disgust Scale[40]. The questionnaire consists of thirty-two items which are separated into eight sub-domains of disgust; food (found unfit for consumption), animals (those usually associated with dirty conditions), body products (most of the bodily solid and fluid extractions, including odours, etc.), extreme and deviant sexual behavior, skin breaches, death and corpses, poor hygiene, and sympathetic magic (stimuli which are non-infectious by themselves, but resemble or come into contact with infectious stimuli).

Evoking a hair in the soup or a worm in a bitten apple will undoubtedly have triggered images in your mind. Alongside emotion and anecdotes, mental images are extremely powerful in orientating our choices and shaping our judgment.[41]

---
[40] Haidt, J., McCauley, C. and Rozin, P., 1994
[41] Haidt, J., McCauley, C. and Rozin, P., 1994

In the Mind's Eye

The late paleontologist and famous science-writer Stephen Jay Gould once said that primates are visual animals, and that the pictures we draw show the things we believe in most deeply, as well as display our conceptual limitations.

Jack Was Killed By a Semi-trailer

Richard Nisbett and Lee D. Ross, Professor of Humanities and Sciences at Stanford University, illustrate the vividness of the image effect by contrasting two descriptions of the same event[42].

Picture: a car accident

In the first description, one learns that *"Jack sustained fatal injuries in an auto accident"*. This description of death evoked weaker emotional reactions than the second description that *"Jack was killed by a semi-trailer that rolled over on his car and crushed his skull"*. The two descriptions create very different mental images for the person listening to or reading them, and even though they both describe the unhappy circumstances, they evoke different sets of emotions.

---

[42] Nisbett, R. and Ross, L., 1980

Picture: a bus accident somewhere in the world

Images, Words, and Emotions

 A Small Dose of Theory On Images and Emotions

Recent work in the field of psychology shows how emotions play a crucial role in decision-making. Neuroscience specialists such as Antoine Bechara, Professor of Psychology and Neuroscience, Hanna Damasio, Professor of Psychology and Neurology, and Antonio R. Damasio, Professor of Neuroscience—all three at the University of Southern California–maintain that these direct emotional influences can cause people to veer away from evaluating a situation on a more intellectual level[43]. They argue that thought is constructed largely of images, which they consider to include both perceptual and symbolic representations. A lifetime of learning results in these images becoming associated with positive and negative feelings which are in turn associated with physical states. Negative associations sound like mental alarms and positive images create positive associations.

---

[43] Bechara, A., Damasio, H. and Damasio, A. R., 2000

Picture: a strange piece of art in the University of Tandil, Argentina

Emotions are notoriously difficult to verbalize. They operate in some psychic and neural space that is not readily accessed from consciousness.
Joseph LeDoux[44]

 Try the Celebrity List Test

Here's a fun experiment that you can try with your friends. Give them the following list of people, with equal numbers of males and females, where the females are famous (movie or TV stars, etc.), while the males are not.[45]

Jodie Foster
James Smith
Eva Mendes
John Murphy
Halle Berry
Robert Jones
Sharon Stone
Michael Brown
Demy Moore
William Davis
Jennifer Aniston
David Miller
Charlize Theron
Richard Wilson
Eva Longoria
Charles Anderson

---
[44] LeDoux, J., 1998
[45] Kahneman, D. and Tversky, A., 1974

After your friends have looked at the list for 10 s, take it away and ask them to estimate the proportion of men to women. Most will estimate that more than half the people on the list are women (when in fact the numbers are equal). The reason why we tend to overestimate the number of women is because all the women's names are famous (at least to American people, and in 2015), while all the men's names are anonymous.

When Detroit Looms Larger Than Michigan

Consider the question: "How many murders were there last year in Detroit/Michigan?" Although, logically, there cannot be more murders in Detroit than in Michigan, because Michigan contains Detroit, the word "Detroit" evokes a more violent image than the word "Michigan" (except of course for people who immediately think of Detroit when Michigan is mentioned).

If people use an impression of violence as a mental shortcut and forget to consider that Detroit is a city in Michigan, their estimates of murders in the city may exceed their estimates for the State, which clearly is not logical. In a large sample of students at the University of Arizona, this hypothesis was confirmed: the median estimate of the number of murders was 200 for Detroit and 100 for Michigan.[46]

The Larger Bowl Looked More Inviting (Though It Offered Fewer Prospects)

In 1994, Veronika Denes-Raj and Seymour Epstein of the Department of Psychology at the University of Massachusetts in Amherst reported on the results of a paradoxical experiment in their paper *Conflict between Intuitive and Rational Processing: When People Behave against Their Better Judgment*.[47] Participants were offered a chance to win a prize if they could draw a red jelly bean from one of two bowls. One bowl contained a larger total number, but a smaller proportion, of red beans (e.g., 9 out of 100 or 9 %). The other bowl had fewer red beans with a higher probability of winning. Denes-Raj and Epstein made the striking observation that participants often elected to draw from a bowl containing a greater absolute number, but a smaller proportion of red beans, rather than from a bowl with fewer red beans but a higher proportion!

Interestingly, many participants reported experiencing a conflict between what they objectively knew were the better odds and how they felt about the

---

[46] Goldstein, D. G. and Gigerenzer, G., 2002
[47] Denes-Raj, V. and Epstein, S., 1994

larger bowl which "looked more inviting". As Denes-Raj and Epstein put it, these participants felt they had a better chance when there were more red beans.

In a second experiment, people tended to prefer 7/100 over 1/10 and an astonishing 34 % of the subjects preferred to draw from the large bowl even when it offered only half the chance (5 %) of winning as the small bowl (10 %).

Picture: what case would you choose if you had to maximize the chance of selecting a red item with your eyes closed?

Media Coverage of Epidemics Calls Images to Mind

Authors such as Helene Joffe, Psychology Professor at University College, London, and Georgina Haarhoff[48], from the Centre for Family Research, University of Cambridge, have analysed how mental images of epidemic diseases or of nuclear disasters, potential or realised, are often at variance with technical assessments from experts in the field.

Helene Joffe analysed media coverage in relation to how people perceive risk on a daily basis. Depending on the visual imagery presented to them by their news source, they feel differently about what has happened.

Joffe reports a study by Jenny Kitzinger, Professor at the Cardiff School of Journalism, Media and Cultural Studies. Kitzinger analyzed how the British reacted to media coverage of AIDS.[49] At that time, treatment for the disease was limited, and it was spreading rapidly through at-risk groups. When presented with information by the media, people are deeply influenced by what they already know or, more to the point, what they *think* they know. By and large, the man in the (British) street saw Africa as the cradle of AIDS, because the media had given him this impression.

---

[48] Joffe, H. and Haarhoff, G., 2002
[49] Joffe, H., 2008

According to Kitzinger, the UK population accepted this hypothesis readily because, to them, it seemed to confirm beliefs they already held about the "black continent" with its periodic devastating famines, disease, and sexuality that people believed to be primitive and perverse. British people found it easy to imagine apocalyptic scenes in Africa that would have been unimaginable in the UK or closer at hand because the mental images they already had of the continent seemed to be compatible with it as a place from which awful diseases could spread.

Ebola Virus Imagery Held Off the 1995 Epidemics In a Science Fiction Limbo

In 2002, Helene Joffe and Georgina Haarhoff[50] explored how British tabloids communicated news about a newly encountered illness: the Ebola virus. Ebola is named after the Ebola River in the Democratic Republic of Congo, central Africa, where the first known outbreak of this disease occurred in 1976. Ebola, from the family of filoviruses, is transmitted via blood. The onset of symptoms is fast: from shivering, high temperature, and headache to death, which tends to occur in the second week of infection.

In the 1995 Wolfgang Petersen movie, *Outbreak*, Dustin Hoffman plays a scientist sent to an isolated African village to study the virus. Hoffman is dressed like an astronaut in a suit and helmet as a protection from the virus while he is investigating the village. Such an impressive safety suit, reminiscent of those used in a space mission, was a confirmation to the general public of one of their preconceived beliefs about Africa, namely, that it is just like another planet, totally different from ours. Therefore, after seeing the movie, one might well feel that nothing that happened in Africa could be of any concern to people in the West. Science fiction, horror, and symbolism associated with Africa fed the sense that Ebola was not "real" for its British audience. It seems too awful to be possible, more like something from television or the cinema than a real-life threat.

The 2014 and 2015 epidemics—which started in Sierra Leone and Liberia, and spread dramatically across several countries in West Africa—got much more attention from American and European citizens, as the disease contaminated thousands of people and, in particular, even highly protected staff from NGOs. The terrifying prospect that there is no cure, and the news that it has reached America and Europe, makes it more vivid to the Western audience.

---

[50] Joffe, H. and Haarhoff, G., 2002

According to Joffe, this may support the claim that imagery exerts a "positioning" power in terms of people's representations of the media. Audience and media strength do not therefore account for the whole explanation of consumer or spectator behavior. Also important is the way the message is positioned regarding cultural and moral beliefs and stereotypes (a fact well known in advertising).

Similarly, when people are presented with terrifying images showing the results of genetic modification, their otherwise negative reactions to GMOs are strengthened.

Lies, Damn Lies, and Statistics

Images and feelings exert a considerable influence that affects the way we deal with figures and numbers. There is even a debate among scholars as to whether or not our minds are designed to handle statistics and probabilistic reasoning.

Tom Is Tall Because He Is Heavy

Daniel Kahneman and Amos Tversky[51] consider the choice between the two assumptions:

- Tom is heavy because he is tall,
- Tom is tall because he is heavy.

The former is preferred over the latter, perhaps because the prototype of the tall man is that of a heavy man. However, such an inference cannot be justified on statistical grounds.

Describing the relationships between causes and effects, Daniel Kahneman and Dale T. Miller, Professor of Psychology at Stanford University, noted that a child may be described as "big for her age" but not "young for her size".[52]

Adding a Small Loss Makes the Bet More Attractive!

Paul Slovic, Melissa L. Finucane, Senior Behavioral and Social Scientist at RAND Corporation, Pittsburgh PA, Ellen Peters, Professor of Psychology at

---

[51] Kahneman D. and Tversky A., 1974
[52] Kahneman, D. and Miller, D. T.,1986

The Ohio State University and Donald G. MacGregor report a puzzling experiment in which a bet becomes more attractive when a small loss is added.[53]

The idea is that you offer a gamble to someone consisting in having seven chances out of thirty-six possibilities to win nine dollars. There is an opaque bowl with seven red balls and twenty-nine others. Each time you pick out a red ball you get nine dollars, but you have to pay a dollar each time you play.

You ask the potential gambler what he thinks of the deal; most won't find it all that great. When people are asked what they find attractive about the gamble they tend to rate it at an average of 9.4 out of 20 points, with the best deal being twenty and the worst zero.

When the gamble is presented with the same odds, seven chances out of thirty-six to win nine dollars, but with the additional risk of losing five cents every time you lose, most people find it much more appealing!

Presented with this possibility, potential gamblers rated it at 15.9 on the 20 scale. If the rational mind was at the heart of the matter the first game should seem better; the emotional mind, however, is more attracted to the second game. Why?

These apparently odd findings can be explained in terms of feeling. People who are very numerate and think about the numbers more than the less numerate are likely to have a more precise emotional reaction when they consider issues of probability.

In contrast, applying a monetary value to the scale makes it less precise because there is no way to determine whether $ 9 is good or bad, attractive or unattractive. Thus, the impression of the gamble offering a win of $ 9 and no losing payoff is dominated by the rather unattractive impression produced by the 7/36 probability of winning. However, adding a very small loss to the payoff puts the $ 9 payoff in perspective (i.e., it makes the payoff clearer and more emotionally available) and gives it meaning, because there is now something to compare with the nine dollar win.

The combination of a possible $ 9 gain and a possible 5ct loss has a very attractive win/lose ratio, making it much more attractive than if no compare-and-contrast was possible. Discussing the possibility of winning $ 9 without mentioning a loss carries less weight.

Distinction Bias

People tend to view two options as more dissimilar when evaluating them simultaneously than when evaluating them separately. For example, when televisions are displayed next to each other on the sales floor, the difference in

---

[53] Slovic, P., Finucane, M. L., Peters E. and MacGregor, D. G., 2002

quality between two very similar, high quality televisions may appear great. A consumer may pay a much higher price for the higher-quality television, even though the difference in quality is imperceptible when the televisions are viewed in isolation. Because the consumer is likely to watch only one television at a time, the cheaper television would have provided a similar experience at a lower cost. This is a widespread selling technique. Just think of supermarket shelves organized by product categories: dairy products, bakery, non-alcoholic drinks, breakfast cereals…

Douglas H. Wedell and Elaine M. Santoyo, both from the University of South Carolina, and Jonathan C. Pettibone, from the Southern Illinois University in Edwardsville, put forward the concept of *"judgmental relativity"* applied to self-perception and the shaping of ideals.[54] Women tend to idealize thinness. Their experiment aimed to test the relationship between the idealization of silhouette thinness by women and their exposure to fashion advertisements showing slim top models. The experiment revealed that the same female figure could be judged wide and unpleasant in one context, and narrow and pleasant in another. The fact of being exposed to different silhouette contexts altered body shape ideals. "*The mere exposure to a skewed set of extremely thin human figures leads to the adoption of a thin ideal for evaluating body images.*"

Evaluations of available alternatives and the choices that people subsequently make are context-dependent. Here thin figures were preferred after women had been exposed to a series of thin figures.

The Gambler's Fallacy (Gamblers Yearn For the Good Outcome That Will Offset All the Bad Ones)

As English writer Samuel Johnson put it, second marriages are the triumph of hope over experience. At the roulette game for instance, the hope that a black should soon follow a long series of reds is known as the *"gambler's fallacy"*. Indeed, the sequence should look more like one would (erroneously) expect a random sequence to be, and should thus be more representative of chance.

Chance is commonly viewed as a self-correcting process where a deviation in one direction induces a deviation in the opposite direction to restore the equilibrium. The so-called "gambler's fallacy" induces people to interpret random processes—such as tossing coins a large number of times—as having self-correcting properties, contradicting the principle of independence between random outcomes. A comprehensive academic paper[55] relates several

---

[54] Wedell, D. H., Santoyo, E. M. and Pettibone, J. C., 2005
[55] Dohmen, T. F., Falk, A., Hufman, D., Marklein, F. and Sunde, U., 2009

studies (such as one by Charles T. Clotfelter and Philip J. Cook[56]) which reveal that lottery players believe in the gambler's fallacy. The Maryland State Lottery outcomes show that, in the days after a winning number has been drawn, the betting on this particular number drops significantly.

Similarly, Rachel Croson and James Sundali[57] demonstrated the prevalence of the gambler's fallacy in the betting behavior of roulette players in a casino in Reno, Nevada. The authors observed that, after a long streak of the same outcome—for instance, red comes up five times in a row—players put significantly more bets on black.

This fallacy is not unique to gamblers; we have all known parents desperate to have a girl after having had several boys, or vice versa assuming that nature "owes" them a child of the other sex.

The Base Rate Fallacy

Kahneman and Tversky carried out an intriguing test.[58]

 Try Testing Yourself!

The following test illustrates how in many cases we are not fully at ease even with rather simple figures.
- On average in the population, 50% of the babies are boys.
- One town has two hospitals.
- In the larger, 45 babies are born each day.
- In the smaller, 23 babies are born each day.
- Over 1 year we measure the number of days when more than 60% of babies were boys.
- Which hospital recorded the larger number of such days?

The correct answer is the smaller hospital. The bigger the hospital, the more births are recorded, and the higher the probability that they conform to the general population average. The smaller the hospital, the higher the probability that the birth distribution departs from the average recorded in a big sample size.

---

[56] Clotfelter, C. T. and Cook, P. J., 1993
[57] Croson, R. and Sundali, J., 2005
[58] Kahneman, D. and Tversky, A., 1974

Pill Scare

Elke Kurz-Milcke, Gerd Gigerenzer, and Laura Martignon, from Ludwigsburg University of Education, address the question of how the public could better understand risks, given the difficulty in handling statistical results.[59] What has been called the "pill scare" shows how people behave against their own interests when they don't have a correct understanding of statistical information. These authors recall that women have gone through many concerns relating to the contraceptive pill since it was introduced in the 1960s. Unfortunate presentation of statistics can provoke a dramatic pill scare chain reaction. In the mid-1990s, the British press reported the results of a study that women who took the contraceptive pill increased their risk of thromboembolism by 100%. Thromboembolism means blockage of a blood vessel by a clot, and it is a condition that can lead to fatal strokes. Hearing the bad news, thousands of British women panicked and stopped taking the pill, which led to a wave of unwanted pregnancies.

But what did the study *really* show?

In fact, of every 14,000 women who did not take the pill, one had thromboembolism, and out of every 14,000 who took it, this number increased from one to two. That is, the *absolute* risk increase is 1 in 14,000, even though the *relative* increase is 100%. Most of these women, like the majority of people, had never learned the difference between absolute and relative risks, and thus were easily frightened by the report citing a 100% increase in the condition. It has been estimated that more than 10,000 British women had abortions as a consequence of the press release, and who knows how many unwanted pregnancies were brought to term.

Are Enemies At the Door?

Daniel Kahneman, Paul Slovic, and Amos Tversky report the following test by David A. Schum, from Ohio State University, who specialises in the properties, uses, and marshalling of evidence in probabilistic reasoning. The test illustrates how quickly we can jump to a conclusion, short-circuiting the logic of numerical calculation.[60]

- *Imagine you are a military general and concerned that your enemies will invade your country.*
- *It is known to you from past experience that when enemy troops mass at the border, the probability of invasion is 0.75.*
- *But you do not have direct access to this information, so you must rely on reports.*

---

[59] Kurz-Milcke, E., Gigerenzer, G. and Martignon, L., 2008
[60] Schum, D. A., 1980, in Kahneman, Slovic, and Tversky, 1982

- *As it happens, every time your intelligence service tells you that they are massing at the border, they are really there.*
- *What is the probability of invasion?*

Try it out on your friends! Most of them will answer "0.75", making a logical error in assuming that, if the troops are there every time the report says they are there, then it is also true that they are not there if there is no report to say that they are there. In fact, we have no confirmation about the latter hypothesis. Therefore, based on the existing data, we cannot deduce the probability of invasion. Indeed, the probability of invasion is the probability of "invasion and having a report" plus the probability of "invasion and not having a report". But the information provided above did not preclude enemy troops from massing at the border in cases where the intelligence service sends no report to that effect.

Furthermore, the outcome of the task, which is invading a country, may be reinforced by the general's decision to mass his troops at the border, leading to invasion by the enemy.

23 People and 2 Birthdays

Probabilities sometimes defy layman's intuition. In a class reunion of as few as 23 people, there is more than 50 % probability that 2 of them share the same birthday (same day and same month).[61]

When 20 Out of 100 Is Not Equal to 20 %

In a series of experiments by Paul Slovic, Melissa Finucane, Ellen Peters, and Donald G. MacGregor—from Decision Research in Eugene, Oregon—experienced forensic psychologists and psychiatrists were asked to judge the likelihood that a mental patient would commit an act of violence[62]. Unsurprisingly, when clinicians were told that "*20 out of every 100 patients similar to Mr. Jones are estimated to commit an act of violence,*" 41 % would refuse to discharge the patient. But when another group of clinicians was given the risk as "*patients similar to Mr. Jones are estimated to have a 20 % chance of committing an act of violence,*" only 21 % would refuse to discharge the patient, despite the fact that the two sentences contain exactly the same mathematical information!

---

[61] Consider the case where there is no pair sharing the same birthday among 23. The first individual has 365 possibilities of birthday date, the second $365-1$ possibilities…the 23rd has $365-23+1$ possibilities; therefore the probability for this situation is $(365 \times (365-1) \times \ldots \times (365-23+1))/(365 \times 365 \times \ldots \times 365)$. This ratio is close to 0.5

[62] Slovic, P., Finucane, M., Peters, E. and MacGregor, D. G., 2004

While 15%, 0.15, 15/100, and 15 out of 100 (or 150 out of 1,000) have the same mathematical content, they don't "feel" the same, and people tend not to perceive them in the same way. This simple fact has serious implications for society in general.

Rebukes Would Boost Learning, While Praise Backfires… Really?

Daniel Kahneman relates, in his autobiography on the Nobel Prize website[63], a discussion he had on flight training with experienced instructors. The latter noted that praise for an exceptionally smooth landing is typically followed by a poorer landing on the next try, while harsh criticism after a rough landing is usually followed by an improvement on the next try. Instructors concluded that verbal punishments are beneficial.

> *I had the most satisfying Eureka experience of my career while attempting to teach flight instructors that praise is more effective than punishment for promoting skill-learning."*
>
> *When I had finished my enthusiastic speech, one of the most seasoned instructors in the audience raised his hand and made his own short speech, which began by conceding that positive reinforcement might be good for the birds, but went on to deny that it was optimal for flight cadets.*
>
> *He said: "On many occasions I have praised flight cadets for clean execution of some aerobatic maneuver, and in general when they try it again, they do worse. On the other hand, I have often screamed at cadets for bad execution, and in general they do better the next time. So please don't tell us that reinforcement works and punishment does not, because the opposite is the case.*
>
> *This was a joyous moment, in which I understood an important truth about the world: because we tend to reward others when they do well and punish them when they do badly, and because there is regression to the mean, it is part of the human condition that we are statistically punished for rewarding others and rewarded for punishing them."*

So, what is regression to the mean? Suppose the instructor measures the performance of a cadet on a scale from 0 to 20. Suppose the mean grade of that cadet is 10, and that marks are symmetrically distributed around this mean (as with a bell curve). Now, assume that the instructor rates the first execution as 4.8 on the scale, much lower than the average of 10.

It is clear from the figure below that the probability that a performance exceeds 4.8 (the shaded surface to the right, below the curve) is more than the

---

[63] http://www.nobelprize.org/nobel_prizes/economic-sciences/laureates/2002/kahneman-bio.html

probability that a performance falls below 4.8 (the unshaded surface to the left, below the curve). Therefore, it is more likely that the second execution will yield a better performance measure than the first.

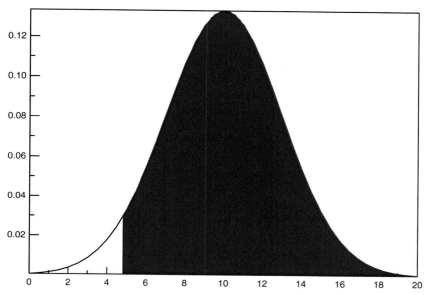

Picture: bell-shaped probability distribution[64] of marks between 0 and 20

That's it! Regression to the mean implies that, statistically, an improvement will usually follow a poor performance and deterioration will usually follow an outstanding performance. But regression to the mean does not contradict independence between outcomes, no more than it supports the gambler's fallacy.

Feelings and Statistics Are Poor Bedfellows

Advertisements for lotteries tend to emphasize successful outcomes.[65] Because of their vividness, images of happy winners are much more memorable and cognitively accessible than vague knowledge about millions of losing tickets.

Intuition Did Not Break the Sound Barrier

We have slogged through many examples of intuitive thinking, many of them appropriate. However, it is clear that rational decision-making is a powerful

---

[64] The authors thank Jean-Philippe Chancelier who made this drawing with the software Scicoslab
[65] Mumpower, J. L., 1988

alternative. We now illustrate the power of mathematical thinking with two stories, in the domain of aeronautical engineering, where intuition was misleading and the use of mathematical models led to great success.

In *The Sound Barrier* (1952), the British movie director David Lean tells the story of the first two jet pilots who tried to break the sound barrier. One protagonist followed his intuition, which was that the aircraft's nose should be dropped. In the event, the plane disintegrated. His gut feeling that the nose needed to drop led to many crashes.

A few years later, pilots and engineers tried to understand the best configuration for safely breaking the sound barrier, and were able to determine precisely what needed to be done.

The Minimum Time Trajectory Defies Intuition

Consider the best way for a supersonic plane interceptor to climb from sea level to a given altitude in least time. Quite naturally, we think that the plane should always ascend.

However, in 1961, Arthur Earl Bryson, Jr, Professor of Engineering Emeritus at Stanford University, and Walter F. Denham used a mathematical algorithm to show that the optimal trajectory has the following features:

> *"the airplane first climbs, then dives accelerating through sonic speed, then pulls up, trading kinetic for potential energy in the earth's gravitational field"* [66]. They stressed that *"These surprising trajectories show the danger of using classical performance methods on high-performance airplanes"*.

Bryson recalls how they calculated *"the minimum time-to-climb path to an altitude of 20 km, Mach 1, and level flight"* and that their surprising trajectory was then tested.

> *The path was tested in January of 1962 at the Patuxent River Naval Air Station. The co-pilot had a card with the optimal Mach number tabulated for every 1000 ft of altitude, which he read off to the pilot as they went through that altitude. The pilot then moved the stick forward or backward to get as close to this Mach number as he could. They got to the desired flight condition in 338 s, where the predicted value was 332 s. This was a substantially shorter time to that flight condition than had been achieved by cut-and-try.*[67]

---

[66] Bryson, A. E. and Denham, W. F., 1962
[67] Bryson Jr., A. E., 1996

7

How to Balance Pros and Cons, and Other Helpful Hints

Shaken by illusions and emotions, will you be more secure on rational grounds? Let us try balancing pros and cons and discover how Charles Darwin reached the decision to marry. And let us learn *"fast and frugal rules"* and discover how to predict divorce from simple clues. You could also become an expert in wine tasting, but could you have finished building the Sydney Opera House earlier?

But for the moment, let's start with anecdotes on how the expert and the layperson make decisions.

Five Anecdotes On How Decisions Are Reached

Here we'll be looking at examples of "fast and frugal rules" à la Gigerenzer.

J. C. Penney Hired Executives "On Salt"

*J. C. Penney, the famous American business man,
had a funny way to hire executives "on salt".
He used to take potential business executives out to lunch.
When they salted their food without so much as tasting it, he didn't hire them,
feeling that this showed a less than inquiring mind.*[1]

US President Ronald Reagan Tested Visitors with Jelly Beans

Ronald Reagan, the late US President, also used a rather idiosyncratic test.

He kept a jar of multi-coloured jelly beans on his desk and offered them to visitors to observe whether they took one, refused a jelly bean, took only beans of all one colour, or a handful, and so forth.[2]

---

[1] Dawes, R. M., 2000
[2] Bar-Hillel, M. and Ben-Shakhar, G., 2000

Economics Nobel Prize Markowitz Did Not Apply His Theory for His Own Investments

Harry Markowitz received his Nobel Prize for an optimal asset allocation method known as mean-variance portfolio (currently advertised by banks worldwide), yet when he made his own investments for retirement, he did not use his optimization method. Instead, he relied on an intuitive heuristic known as 1/N: allocate your money equally to each of N alternatives.[3]

There are times when logic and analytical thinking do not seem to work and the mind takes over and assesses the situation on a less than fully conscious level.

This mismatch between practice and theoretical thought is illustrated by the following urban legend, recounted in academic circles.

C'mon, This Is Serious!

An urban legend has spread among economists:

> An economics professor from Columbia University had an offer from Harvard. He could not make up his mind whether he should accept or reject it and was getting stressed about the decision he was facing.
> A colleague took him aside and said,
> "What is your problem? Just maximise your expected utility![4] You always tell your students to do so."
> Exasperated, the professor responded, *"C'mon, this is serious."*

Garry Trudeau's Lists

In a cartoon,[5] the famous American cartoonist Garry Trudeau imagined two characters, walking past each other in the street. We can read their thoughts as written in two comic strip bubbles. Indeed, we can read in those bubbles how they are deciding whether or not to greet one another. They are doing that by reviewing a lengthy checklist of rational reasons, amusingly employing

---

[3] Gigerenzer, G., 2008
[4] Expected utility is the mainstream economics theory about decision-making under risk
[5] http://www.google.fr/url?sa=t&rct=j&q=&esrc=s&source=web&cd=1&ved=0CCUQFjAA&url= http%3A%2F%2Fyhs.yorktown.org%2Fdownload.axd%3Ffile%3Da7bc8b513eba453e9a8ee865b3a ef737%26dnldType%3DResource&ei=UUC9U8mxGsLQ0QWr4YDQCg&usg=AFQjCNGdnbglS9 -DP7bLqw2ZtcJS72DeYQ&bvm=bv.70138588,d.d2k

a systematic analysis of the risks and risk-mitigating factors implicit in this situation.

What makes this cartoon funny is the fact that we instantly recognize that no one in such a situation would ever be this analytical, even if their life was at stake—or at least not on a rational and conscious level. Instead, most risk analysis is handled quickly and automatically when we reach for our mental Swiss army knife and take out the most appropriate tool.

Piece-of-Cake Algebra Predicts Better Than Personal Judgment

Can number-crunching sometimes provide answers in areas where emotions usually rule? Let's look at some examples from the great and glorious.

How "Unromantically" Did Charles Darwin Decide to Marry

In 1836, Charles Darwin returned to England upon the conclusion of his seminal journey in *The Beagle*. He had crossed oceans, explored wild lands, and made exciting discoveries. As we now know, he had already started a journey that would vastly enrich human learning and completely change the way scientists and ordinary people alike would understand the world around them.

Now he had to make a decision about the most important journey of all: the one from bachelorhood to marriage!

On this particular day in 1838, Darwin was confronted with an intimate personal concern, and a very important decision: should he or should he not marry his charming, intelligent, and cultured cousin, Emma Wedgwood?

Ultimately, Darwin decided to get married (and he and Emma would go on to raise a large family together). In the process of coming to a decision, it seems that he tried to rationalise a choice in which one might have expected a more emotive approach. He applied the power of his considerable intellect to a matter of the heart, favouring thinking over feeling. He drew up a column with the pros and another with the cons, and weighed them somehow.

Picture: to marry or not—2nd Note, MS Dar 210.8:2r, ©Cambridge University Library

Extract of Charles Darwin diary:

Marry
Children (if it pleases God)
Constant companion (and friend in old age) who will feel interested in one.
Object to be beloved and played with,
Better than a dog anyhow.
(…)
Not marry.
(…)
No children, (no second life), no one to care for one in old age.
(…)
Freedom to go where one likes, choice of Society and little of it.
Conversation of clever men at clubs.
Not forced to visit relatives, and to bend in every trifle, to have the expense and anxiety of children perhaps quarrelling.
Loss of time. Cannot read in the Evening, fatness and idleness.
(…).

(The full text can be found on http://www.darwinproject.ac.uk/darwins-notes-on-marriage)

Darwin's diary provides us with insights into the profound duality of humankind. On the one hand, we are visceral-feeling beings that often react intuitively, without realising why we do the things we do, or even thinking about it that much. When we try to rationalise and analyse our more instinctive reactions, we often fail. On the other hand, we can also approach life from a more cognitive, rational standpoint, favouring thinking over feeling. In practice, however, we are not always fully aware of which aspect of ourselves, thinking or feelings, is in charge when we make decisions.

Benjamin Franklin Advocates "Moral Algebra" to Weigh Up Pros and Cons

In the following letter to Joseph Priestley, Benjamin Franklin explains the pros and cons method, which he calls "Moral or Prudential Algebra". Hereafter you will find several significant extracts.[6]

> *To Joseph Priestley*
> *Dear Sir, London Sept. 19. 1772 (…)*
> *…my Way is, to divide half a Sheet of Paper by a Line into two Columns, writing over the one Pro, and over the other Con. Then during three or four Days Consideration I put down under the different Heads short Hints of the different Motives that at different Times occur to me for or against the Measure. When I have thus got them all together in one View, I endeavour to estimate their respective Weights; and where I find two, one on each side, that seem equal, I strike them both out: If I find a Reason pro equal to some two Reasons con, I strike out the three. If I judge some two Reasons con equal to some three Reasons pro, I strike out the five; and thus proceeding I find at length where the Balance lies; (…)*

Then, Franklin advises Priestley to consider separately those reasons that cannot be easily weighted algebraically. When the whole array of positive and negative reasons are so displayed in front of him, Franklin takes this to be a good way of avoiding irrational decisions.

Paul Meehl's Review of Evidence

Two American psychologists showed that Franklin's approach proved efficient in a range of settings, including ratings within university faculties and interviewing candidates. Robyn Mason Dawes followed the path of Paul Everett Meehl, who claimed that the holistic judgment, when applied to clinical treatment or student selection, was pretty poor compared to more systematic

---

[6] The full text of the letter can be found for instance on http://franklinpapers.org/franklin/yale?vol=31&page=455a

ways. Robyn Dawes appraised Franklin's prudential algebra as an efficient *"weighted average of reasons for or against doing something"*.[7]

Literature searches carried out over decades by Meehl and Dawes failed to show any non-ambiguous studies in which clinical judgment had been shown to be superior to statistical prediction. Dawes provides the example of the doctor who spends a lot of time face-to-face with his patients, but who reads no medical journals, orders few or no tests, and *"grieves at the funerals"*.

One of the authors of *The Biased Mind* knew one of these very empathic country doctors. This doctor felt so close to him for years that radiological examinations or scanners seemed useless. That was until a test at a hospital downtown revealed that the seemingly harmless symptom (the one he had been living with for months) might well be more severe than "diagnosed".

To no avail, decade long reviews of evidence back up number crunching as more effective than personal judgment in medical prediction.

In 1954, Paul Meehl published *Clinical vs. Statistical Prediction: A Theoretical Analysis and a Review of the Evidence*,[8] claiming that mechanical data analysis performed better in terms of predicting behavior than subjective, personal methods. Looking at the field of medicine, he argued that using mechanical methods would provide the most reliable way to come to a decision about patient treatment and prognosis.

Dry Linear Models Could Usefully Complement Smart Expert Categorization

> People tend to be much more confident than they should be in terms of their judgements. However, experts in any given field tend to be "much better at selecting and coding information than they are at integrating it."[9]
> Robyn Dawes

We need experts to identify the few relevant features of a complex situation. But it is a different matter altogether to use those insights in order to assess and decide. For the latter task, we would do better to rely on linear models (adding pros and cons) than on experts. A good example of this discrepancy between expert and model efficiency is reported by Robyn Dawes in the case of the diagnosis of Hodgkin's disease.[10] That study showed that overall subjective assessment made by expert doctors on the severity of the disease among

---

[7] Dawes, R. M., 1979
[8] Meehl, P., 2013 (orig. 1954)
[9] Dawes, R. M., 2000
[10] Dawes, R. M., 1979

193 patients was not as good as a multiple regression model (using the same expert's variables) when it came to predicting their survival time.

Doctors' variables proved to be of real significance: experts know what to look for. However, the most efficient integration of the variables comes out of a regression model.

The moral of the story is that you need experts for what they are good at—extracting relevant features—and linear models for proper assessment by aggregating marks on a scale.

Another example of the power of crude linear models comes from the Denver City Police.[11] They had to choose a better bullet calibre because the one they used was not very efficient. Everyone from politicians to policemen had an opinion. They finally came up with a simple analysis based on weight and parameters that indicated that they should use a bullet model that nobody had ever considered.

Marks Do Better Than Interviews (Though Both Are Weak Predictors In the Absolute)

Robyn Dawes describes a personal experience.[12]

> *When I was at the Los Angeles Renaissance Fair last summer, I overheard a young woman complain that it was "horribly unfair" that she had been rejected by the Psychology Department at the University of California, Santa Barbara, on the basis of mere numbers, without even an interview.*
> *"How can they possibly tell what I'm like?"*
> *The answer is that they can't. Nor could they with an interview.*

Dawes mentions E. Lowell Kelly's work on the evaluation of psychological assessment techniques, hinting at the poor information generated by unstructured interviews (beyond finding out about past behavior, and establishing whether the interviewer likes the interviewee, which may be important). Many insist that it's *"dehumanising"* to make important choices without interviewing someone, but adding a few numbers that represent the values of relevant features has relatively more predictive power than interviews, although both remain weak in absolute terms.

Dawes reports the reluctance of an academic dean of graduate admissions to rely on a linear model rather than on the judgement of a committee. The dean maintained that the 0.4 correlation between academic marks and future

---
[11] Dawes, R. M., 1979
[12] Dawes, R. M., 1979

faculty ratings although doubling the 0.2 correlation obtained on the basis of the admissions committee's judgment, was still poor. Dawes answered: *"I can only point out that 16% of the variance is better than 4% of the variance"*. (Indeed, a correlation of 0.4 corresponds to $(0.4)^2 = 0.16 = 16\%$ of the variance, while 0.2 accounts for $(0.2)^2 = 0.04 = 4\%$.)

Put another way, future faculty ratings of any individual remain largely unpredictable (84% of the variance). However, a small but significant improvement can be obtained by a linear model, over subjective judgements.

So even though marks show relatively weak predictive or explanatory power (in absolute terms), marks do better than interviews when predicting future academic scores. Indeed, marks are themselves the aggregate of numerous evaluations which, by averaging out fluctuations, can reveal a trend.

By contrast, subjective evaluations open the door to highlighting some features while neglecting others in the interview process. As Dawes[13] remarks, when admissions officers observe good academic results, they can claim that their subjective judgment was right in the first place, neglecting the possibility that good results might actually be put down to the quality of the program. Subjective assessments can also lead to reinforcement, as in the situation where *"a waiter who believes that particular people at the table are poor tippers may be less attentive than usual and receive a smaller tip"*.

An Aside On Assessments and Admissions

Imagine you are the boss and you want to select the best engineer for the task. But what do you mean by "the best"? The best problem-solver? The best at writing report? At animating the team? The best academic record? Here you are confronted with two issues: how to define relevant criteria and how to aggregate those criteria. What the literature suggests is that, when we have to make a choice, we might be good at grasping the relevant categories, the right questions to ask, the main points, etc., but when it comes to aggregating all the relevant factors, we would be better off entrusting that task to crude calculation.

Now, let us turn to another interesting contribution to analysis of the selection process by Eldar Shafir, Professor of Psychology and Public Affairs at Princeton University. Shafir distinguishes two kinds of attitude depending on whether we *"decide to first eliminate those options that we do not want"*

---

[13] Dawes, R. M., 1979

(a selection by rejection), or whether we decide to select the object/subject by choosing the features we like.[14]

Suppose you are part of a recruitment panel/jury, with the mandate to retain only one candidate out of ten. All candidates have sent their resumé. Shafir relates an experiment by Vandra L. Huber, Margaret A. Neale, and Gregory B. Northcraft[15] in which more candidates were chosen when the panel focused on rejecting those who were unsuitable for the job, than when they chose the ones that seemed to be ideal.

If the recruiter is looking to maximize the number of positive attributes and to get "the one" out of hundreds candidates, could it be a good idea to design a procedure in two stages: a first stage during which candidates are shortlisted on the grounds of strengths, good CVs, and so forth, (thus narrowing down the number of forthcoming interviews), followed by a second stage in which they are subjected to, for example, tests and interviews?

The second stage would then consist in interviewing those on the short list of candidates preselected on the grounds of positive attributes.

It would subsequently be easier to identify negative qualities and to reject those candidates preselected on this basis.

 A Helpful Tip On How to Make Short Lists and Recruit

To sum-up, we would recommend a two-stage recruiting process. During the first stage you want to come out with a small number of preselected candidates for subsequent interview. A stricter "positive picking" procedure is likely to leave you with fewer candidates in the second stage of the process. For instance, you will look for specific strengths associated with the position. During the second stage, since the remaining candidates were preselected according to their perceived qualities for the position, it will not be easy to distinguish them according to their positive attributes. You liked them all, that's why you preselected them. We suggest that you arrive at the happy winner by eliminating the other candidates according to their perceived weaknesses.

---

[14] Shafir, E., 1993
[15] Huber, V. L., Neale, M. A. and Northcraft, G. B., 1987

Crude Arithmetic Can Predict Marital Happiness

A substantial academic literature deals with the identification of divorce predictors. It points out that the way couples tell the story of their relationship—for instance using "*we*" rather than "*I*"—can predict the likelihood of marital stability or divorce.[16]

Psychologists Sybil Carrère, Kim T. Buehlman, John M. Gottman, all three from the University of Washington, James A. Coan, University of Arizona, and Lionel Ruckstuhl, University of Nevada, Reno, tested the suggestion that the first few minutes of an arguing couple could predict divorce. "*(…) It was possible to predict marital outcome over a 6-year period using just the first 3 min of data for both husbands and wives.*"

From this amazing study, we thus learn that the relevant and sufficient information about divorce prediction is there, in the first 3 min of a 15 min discussion on marital conflict. This study contributes to the identification of simple rules or clues that have predictive powers for understanding relationships as complex as marriage.

A simple linear model, adding up advantages and disadvantages and calculating the difference between the two, could well be a powerful tool for predicting one's marital future!

It was shown[17] that one could simply count the number of episodes of lovemaking and subtract this from the number of fights. When there was more love-making than fighting, the marriage was going to be a happy one. When there was more fighting than love-making, couples were better off wrapping things up.

 A Helpful and Crude Tip for Testing Marital Happiness

A simple rule for testing marital happiness was determined at the University of Oregon.[18] Dawes discovered that "*a crude improper linear model*" can judge even something as complex as marital happiness.

The take-home message is a simple one:

> When we love more than we hate, we are happy, and when we hate more than we love, we are miserable.

---

[16] Carrère, S., Buehlman, K. T., Gottman, J. M., Coan, J. A. and Ruckstuhl, L., 2000
[17] Dawes, R. M., 2000
[18] Dawes, R. M., 2000

(In fact, one of the subjects of the research decided to get a divorce after realizing that she was fighting more than she was loving!)

Experts Excel In Extracting Relevant Features and Categorizing

There may be said to be two classes of people in the world; those who constantly divide the people of the world into two classes, and those who do not.
American humorist Robert Benchley[19]

Experts quickly go to the point by extracting relevant information in appraisal or problem-solving.

In his book *59 Seconds: Change Your Life in Under a Minute*,[20] Richard Wiseman tells a very old story, often used to kill time during training courses, in which a man is trying to fix his broken boiler, but fails despite his best efforts. He decides to call in an engineer, who simply gives one gentle tap on the side of the boiler and instantly brings it back to life.

The engineer presents the man with a bill, and the man argues that he should pay only a small fee as the job took the engineer only a few moments. The engineer quietly explains that the man is not paying for the time he took to tap the boiler but rather the years of experience involved in knowing exactly where to tap.

In their book, *De la justification. Les économies de la grandeur*,[21] Luc Boltanski and Laurent Thévenot, both at the École des Hautes Études en Sciences Sociales, Paris, report the following anecdote. A woman approached Picasso in a restaurant and asked him to scribble something on a towel. She claimed she was ready to pay whatever he asked.

Picasso complied and said: "*It will be ten thousand dollars*".
"*But you did it in thirty seconds!*"—answered the amazed woman.
"*No*", said Picasso, "*It took me forty years to get there.*"

Similarly, air pilots are experts to whom we entrust our life. Hoping they will be as quick-witted as the US Airways Flight 1549 pilot, who, in 2009, experienced engine problems just moments after take-off from LaGuardia Airport in New York City. At that point, Captain Chesley Sullenberger made a quick

---

[19] Wiseman, R., 2010
[20] http://quoteinvestigator.com/2014/02/07/two-classes/
[21] Boltanski, C. and Thévenot, L., 1989

decision to land the plane on the Hudson River. Even with some four decades of flying experience, it was highly unlikely that at that point in time, facing the probability that everybody on the plane was going to die, Sullenberger was in the right state of mind to rationalise and perform mathematical calculations of how the odds were stacked. Yet somehow he arrived at a decision that allowed all 155 of his passengers and crew to return to safety.

Intuitive London Magistrates

The pilot had to be quick to react, on the basis of limited information. What if you have plenty information and plenty of time to think about it? Let's look at judges in justice.

A study by John Hutchinson and Gerd Gigerenzer[22] revealed that London magistrates display intuitive decision-making, something that is at odds with protocol. A high proportion of the decisions made by London magistrates, as to whether to grant unconditional bail or to make a punitive order such as custody, is based on intuition.

When magistrates were asked how they made their decisions, their replies were at odds with the protocol, which said that they should take many other cues into account, such as the severity of the crime and whether or not the defendant had a home. It could be that they used simpler mental shortcuts unconsciously, but we do not know what information was considered or how it was processed.

Maybe they extract and select a small number of relevant dimensions. But do judges all agree on the frontiers between categories?

Beyond Words, Do Experts Agree On the Occurrence of Unlikely Earthquakes?

We all tend to categorize. For instance, a lot of people assume that they can recognise a professional by the way he is dressed. In *The Picture of Dorian Gray*, Oscar Wilde writes: *"With an evening coat and a white tie...anybody, even a stock broker, can gain a reputation for being civilized"*. Another way of saying "Clothes make the man"!

In the 1990 paper he co-authored with W. Kip Viscusi, Richard Zeckhauser relates the results to a questionnaire addressed to experts about the magnitude and likelihood of earthquakes in the region of San Francisco.[23] Although the experts converged on the answer given to the question

---

[22] Hutchinson, J. M. C. and Gigerenzer, G., 2005
[23] Zeckhauser, R. and Viscusi, W. K., 2000

> *Is there a reasonable chance of a moderate to large earthquake in the San Francisco Bay area in the near future?,*

they appeared to disagree to a large extent on the definitions given to the words *"reasonable", "moderate", and "near future"*. For instance, understanding of a "reasonable chance" could range from 5 to 92%, and the relevant magnitude of the earthquake was anywhere from 5 to 7.5.

And Zeckhauser added:

> *Because the Richter scale is a log scale, the difference between the magnitudes of these earthquakes is 100 fold. The San Francisco Bay area was defined as within 5 miles of downtown San Francisco at one extreme and over 140 miles in all directions on any of fourteen different fault zones at the other extreme. And as everyone knows, the near future ranges from 3 weeks to 100,000 years.*

Zeckhauser was making the point that terminology such as that used in the example is not effective when it comes to making decisions about policy, while this is indeed how policy issues tend to be discussed. Professionals know that there is often a huge range when it comes to the meaning of qualitative terms for uncertainty such as "a reasonable chance".

Expert Burglars Rely On a Few Selective Cues

Research has explored some unusual areas in which "fast and frugal" decision-making rules apply. Investigators studied the decision-making strategies of burglars who rob houses, and they obtained some surprising results. An original study by Rocio Garcia-Retamero, University of Granada, and Mandeep K. Dhami, University of Middlesex, discovered that expert burglars rely on a few selective cues… we leave the interested reader to check out the results in the original paper.[24]

---

[24] Garcia-Retamero, R. and Dhami, M. K., 2009

Become a Wine Tasting Expert

Picture: a glass of Bordeaux wine with a piece of Parisian *baguette* and a chunk of *Pavé d'Auge* cheese

A few years ago in Bordeaux, one of the authors, then a wine professional, conducted various wine tasting sessions and experiments. Regarding tasting methodologies, he noticed that experts are more refined in their descriptions of categories and in their taste vocabulary than mere amateurs, and furthermore that experts and amateurs do not attach the same importance to broad tasting categories. For instance, experts put more weight on the visual, roughly the same on the nose, and less on the mouth. For any individual, expert or amateur, her overall hedonic judgement is strongly correlated with the assessment obtained by adding all the notes and weighting them accordingly. Overall, non-experts and experts alike display very similar assessments for a given wine, judging it either by hedonic ratings or by a crude linear model.

> **A Helpful Tip On How to (Almost) Become a Wine Tasting Expert**
>
> Try the simple wine tasting questionnaire:
> TASTING INVOLVES 3 STEPS: 1. Appearance; 2. Aroma; 3. Palate
> Q1 How would you rate the appearance of this wine? From outstanding (100) to very poor (0)(Average 50)
> Q2 How would you rate the smell/ aroma of the wine? From outstanding (100) to very poor (0)(Average 50)
> Q3 How would you rate the taste of the wine? From outstanding (100) to very poor (0)(Average 50) And finally:
> Q4 What is your overall opinion of this wine? From outstanding (100) to very poor (0)(Average 50)

And then use this non-expert wine lover's tasting sheet for red wine. You may use it for any red wine, between friends.

Wine sample		eg Red merlot from Bordeaux	
Promise		Coefficient	Intensity from 0 to 100
Visual appearance		0,1	60
Nose		0,2	60
Mouth		0,7	55
Arithmetic mark (max 100)		1	57
Spontaneous hedonic mark (max 100)			54

In the example above, non-experts agreed to consider that mouth judgement would weigh 70% of the overall mark, nose 20% and appearance 10%.

Then compare with this wine experts tasting sheet for the same Merlot:

Promise	Expert criteria	Coefficient	Intensity from 0 to 100
Visual		0,25	54
	Ruby colour		54
Nose		0,25	52
	Intensity		53
	Fruit		50
	Spices		53
	Complexity		53
Mouth		0,5	58
	Soft tanins		68
	Concentration		46
	Sweetness		73
	Balance		54
	Finish		51
Arithmetic mark (max 100)		1	56

Experts use more refined words to express their judgements and choose slightly different weights for visual appearance, nose, and mouth. Although they were able to define more precise categories describing the wines, their overall judgement was close to the one made by the wine lover. Similarly, their spontaneous hedonic overall judgement was close to the mark they obtained through calculation of all subcategory evaluations.

One might think that "experts" are better able than most of us to make rational, thoughtful decisions about the world and the things that happen in it. So can we sit back and assume that the professionals we trust to make decisions about our physical, fiscal, and legal health are better placed than we are? Not always. Just as much as the tribesman who still hunts and gathers, the suit-clad professional is very much a caveman in the modern world.

Overconfident Experts

We are often excessively optimistic about what the outcomes of our tasks will be. We tend to assume that everything will work out as planned and forget to factor in things that might have an impact on what needs to be done.

And with hindsight, we infer our success from our own acts and our failures from those of others, or from some features of the environment. My gains on the stock market are due to my smart analysis and my bold moves but, when I lose, I blame the selfish world of finance.

Even Cold-Blooded Experts Betray Over-Optimism

Paul Slovic, Baruch Fischhoff, and Sarah Lichtenstein state that experts, like laymen, are prone to overconfidence biases once they are forced to go beyond their data and to exercise judgement.[25] Using the example of mechanical engineers dealing with the problem of an engine that fails to start, they noted that even people with 15 years' experience in the field could overlook or misjudge pathways to disaster.

The wreck of the Titanic is a tragic illustration of a bias towards overconfidence: there were not enough lifeboats, because the Titanic was believed to be unsinkable, so people simply did not seriously assess the need for lifeboats.

We have already mentioned self-deception in relation to the Forer effect. Over-optimism may also be a form of self-deception, favored by natural selection.

---

[25] Slovic, P., Fischhoff, B. and Lichtenstein, S., 1982

Watch Out! Better Not Read This Chilling Study Before Embarking On an Airplane

In an amazing study, psychologist John A. Swets wonders how often experts spot cracks in airplane wings.[26]

Picture: I can't see any crack

He obtained some data from 150 metal specimens, some with cracks, some without, and sent it to 17 different air force bases. Technicians were asked to examine images of those specimens and determine which had cracks, and which did not. One would expect the experts to agree unanimously when a wing is cracked and a wing is not. But this was not the case!

Swets observed that experts who studied images in order to detect metal fatigue in aircraft wings tended to come up with very different opinions, and there was no system in place to distinguish which experts were the most efficient. There is a threshold to distinguish a good from a bad wing; some of the experts were very strict with a low threshold, while others were much more relaxed; so answers were different because the crack detection threshold varied from one expert to the next. Acknowledging his powerlessness in trying, for several years, *"to get people to become interested in trying (…) techniques to improve the diagnosis of cracks in airplane wings"*, John A. Swets concluded: *"For

---

[26] Swets, J. A., 2000

the moment, one has to hope that any plane one flies on was checked at a good facility rather than a poor one."

The Monty Hall Problem Baffled Mathematical Experts

The following story is taken from an article in the New York Times;[27] it illustrates how overconfident experts can be. The magazine Parade featured a column by Marilyn vos Savant, who is listed in the Guinness Book of World Records Hall of Fame for *"Highest I.Q."*

 Test the Monty Hall Problem

In September 1990, Vos Savant answered a question sent by a reader that went as follows:

Suppose you're on a game show, and you're given the choice of three doors: Behind one door is a car; behind the others, goats. You pick a door, say No. 1, and the host, who knows what's behind the other doors, opens another door, say No. 3, which has a goat. He then says to you, 'Do you want to pick door No. 2?'

Try answering the question yourself!

Vos Savant answered that the contestant should switch doors, although the car could have been placed behind any given door initially. Since her answer, she has received about 10,000 letters, including many from people claiming to be experts in mathematics, stating their disagreement with her conclusion and even trying to make a fool out of her; some even said *"You* are the goat!" However, it turned out that she was perfectly correct. To settle the matter, a quiz show host, Monty Hall, decided to host an experiment in his home.

In his dining room, Mr. Hall put three miniature cardboard doors on a table and represented the car with an ignition key. The goats were played by a pack of raisins and a roll of Life Savers. After 30 independent attempts were made to win the car, the evidence showed very clearly that switching dramatically increased the odds of winning.

Let's imagine the same game but, this time, with 99 goats behind 100 doors and 1 car behind one of those 100 doors. You select one door (without

---
[27] Tierney, J., 1991

opening it) and, after your move, the host opens 98 doors revealing 98 goats behind. I suspect that, like us, you would change your mind and opt for the door the host did not open, rather than stick to your original choice!

Although ultimately Vos Savant was proven to be right mathematically, very few of the mathematicians who had previously denounced her conclusion wrote to her newspaper column to withdraw their previous remarks.

In the 2008 movie *21*, a similar question is asked of the character Ben-Campbell (played by JimSturgess) by Professor Micky Rosa (played by Kevin Spacey). Ben answers correctly, thereby convincing Professor Rosa that Ben would be a good addition to their "card counting team", because he could answer such a difficult and puzzling question.

The Planning Fallacy, Or How We Deceive Ourselves

Hegel was right when he said that we learn from history that man can never learn anything from history.
George Bernard Shaw

Kahneman and Tversky[28] recall that scientists and writers *"are notoriously prone to underestimate the time required to complete a project, even when they have considerable experience of past failures to live up to planned schedules"*. Discrepancies between estimates and outcome may come from ignorance of past outcomes in similar cases as well as from wishful thinking.

Today the Sydney Opera House is famous for its unusual, very beautiful, architectural form. However, it is also very famous for being one of the world's greatest planning disasters! In 1957, estimates for how much the building would cost to build came in at $ 7 million, with the doors due to be opened in 1963. In the event, the building was finally opened in 1973 and it cost the modest sum of $ 102 million!

Another striking and more recent example of expert overconfidence is provided by the Finnish Olkiluoto 3 nuclear reactor. The construction of the EPR reactor began in 2005 and was originally scheduled for completion in 2009. Early in 2012, TeollisuudenVoima (TVO) said the start might be delayed until 2016. On February 28, 2014, Finland's TVO said it could be delayed until at least 2018 as work had slowed. Along with the 9 year delay, the overall budget has been estimated at 7.4 billion €, twice the original budget.

---

[28] Kahneman, D. and Tversky, A., 1982

8
I Frame, You're Framed

And now it's time to bring together the fruits of our mind tour, to digest that abundant food for thought, and make the most of it. Do you feel ready to connect, communicate, and interact with other biased minds around you? You might find some help with tips on message framing. For example, you may work out why the point-down red triangle is such an efficient warning, design efficient nudges (or hints), decipher questionnaires, and perform better than Harvard medical students in a test. Eventually, you will mobilize metaphors and the theory of the embodied mind, and meet smiley's ancestors along the way.

Have a safe journey home.

Pictures: several warnings with different colours, shapes, and texts can convey different meanings. Respectively "caution", "information" and… the entrance to a zoo.

Why Don't We Stop At a Green Light?

Let's see how to communicate warning by colours and signs. Is any colour instinctively and universally associated with danger and risk? It seems that a very small number of colours are.

According to Michael S. Wogalter, Professor Emeritus of Human Factors & Ergonomics, and Christopher B. Mayhorn, Department of Psychology, both at North Carolina State University, red *"is typically perceived to be more hazardous or urgent than other colours"*.[1] Similarly, it has been observed that black, red, and orange are the "top three" colours associated with hazards in both China and the USA.[2]

Many findings point to the fact that red may be the only appropriate colour to use in warnings to denote the presence of risk. Curt C. Braun and N. Clayton Silver[3] showed that the colour red was associated with the highest level of hazard, followed by orange, black, green, and blue.

A Small Dose of Theory On the Color Red

In The Adapted Mind[4], cognitive scientist Roger N. Shepard, Professor Emeritus of Social Science at Stanford University, recalls that:

"The overall (400–700 nm) range of spectral sensitivity of the human eye has long been regarded as an evolutionary accommodation to the range of solar wavelengths that reach us through the earth's atmosphere", and that:
"most languages have terms that native speakers apply to colors in the very same regions of color space for which we use the words "red" and "green"".

As red marks the sunset and the coming of darkness, the color red certainly signaled the need for attentiveness. With poor nocturnal vision, humans had to look for a safe place for the night to shelter from predators. This might be one of the reasons why red is the color that most strongly evokes risk.

---

[1] Wogalter, M. S. and Mayhorn, C. B., 2005
[2] Lesch, M. F., Rau, P. P., Zhao, Z. and Liu, C., 2009
[3] Braun, C. C. and Silver, N. C., 1995
[4] Shepard, N., 1992

Picture: a nice red sunset, time to make it to a refuge, stop some activity, or go to sleep

In general, shapes that appear unstable are preferred as warnings.[5] What happens when instability is presented in association with the red colour?

Picture: which of the two warnings says danger out loud?

The point-down triangle is the most preferred warning shape, followed by the diamond and then the octagon. The point-down triangle evokes instability,

---
[5] Lesch, M. F., Rau, P. P., Zhao, Z. and Liu, C., 2009

calling to mind our innate sense of Earth's gravity. When we see the triangle standing on a point, we feel that gravity will make it fall to the right or the left.

Of course this could be like deciding which appeared first between "chicken and egg": we might wonder whether these observations do not merely reflect our repeated exposure to road signs.

Several studies testing the efficiency of different kinds of warnings show that some combinations of words, colours, and shapes are more efficient than colours or shapes or words of warning alone. Rui-Feng Yu, from Tsinghua University, Alan H. S. Chan, from the City University of Hong Kong, and Gavriel Salvendy, from Purdue University, consider the following six words to be particularly useful in this respect: *danger, warning, caution, alert, prevent,* and *notice*[6]. These words were combined with five different shapes: rectangle, diamond, circle, point-up triangle, and point-down triangle. In general, all six words produced higher ratings when combined with the inverted triangle and the lowest ratings when combined with the rectangle.

Picture: beach warnings without comments: everything is forbidden… have a nice day!

Picture: is this African resort warning efficient without any symbol attached?

---

[6] Yu, R. F., Chan, A. H. S. and Salvendy, G., 2004

Pictures: a beach sign with the warning "No diving" is compared with the symbol without cautionary words (Lake Geneva, Switzerland)

According to Michael S. Wogalter, Michael J. Kalsher, from Rensselaer Polytechnic Institute, and Raheel Rashid[7] regarding the comparative efficiency of warnings, *"deadly"* or *"danger"* convey the strongest intensity, and are even more effective deterrents when associated with the nature of the threat and the indication of the source. The longest and most specific warning is the most efficient, as in the case of *"US Federal Government Warning"* compared to *"Warning"*, or *"Danger, According to Government Regulations, Toxic Waste in this Room"*, as opposed to *"Danger, Toxic Waste"*, or *"Danger, Do Not Trespass, Dangerous Dog"*, as compared to *"Warning, Dangerous Dog"*. The addition of text indicating that the information comes from, for example, the *"US Federal Government"* makes people sit up and pay attention.

Picture: no weapons allowed in the mall

---

[7] Wogalter, M. S., Kalsher, M. J. and Rashid, R., 1999

Picture: a warning worth remembering

Emphasizing Losses, Gains, and Anchors

How best can we communicate with our biased mind on the important matter of health? How can a message be communicated more effectively to a population at risk of cancer? Let us look at an example which deals with ways of communicating prevention issues, and which can be readily extended to contexts other than health.

Changing Two Words Makes a Striking Difference

Breast cancer is a common form of the disease, affecting many women, often fatally, although not as often as before. Early detection massively increases the likelihood of a positive outcome, and this is facilitated by women spending just 5 min a month examining their own breasts. Just *five*. That's less time than most people spend over their morning coffee every day. Despite this, many women never or almost never examine their breasts, leaving them vulnerable to undetected cancers, which can grow and metastasise until they pose a serious risk to life.

Beth E. Meyerowitz, Professor of Psychology at University of Southern California, and social psychologist Shelly Chaiken made a substantial contribution to the literature on framing with a study on how best to design a health pamphlet[8].

---

[8] Meyerowitz, B. E. and Chaiken, S., 1987

They compared the following two messages in terms of efficiency. Notice that they differ in only two "slots":

You can gain several potential health benefits by spending only five minutes each month doing breast self-examination.
You can lose several potential health benefits by failing to spend only five minutes each month doing breast self-examination.

In third position, *"gain"* has been replaced by *"lose"* and, in ninth, *"spending"* by *"failing to spend"*. That's all. But that apparently minor change induces significant consequences!

Indeed, Meyerowitz and Chaiken found that those who read the second pamphlet—framed in terms of all the things that one can lose if breast cancer is allowed to develop—manifested more positive breast self-examination attitudes, intentions, and behaviors (57 % as opposed to 38 % at the 4-month follow-up) than those who read the first pamphlet, framed in terms of gains.

Perhaps counter-intuitively, by focusing on the negative, it becomes easier to encourage women to engage in positive behavior when it comes to safeguarding their own future health and well-being. This exploits the loss aversion behavioral mechanism, demonstrated by Kahneman and Tversky with their Prospect Theory.

This is what we refer to when we say "framing"; how information is presented clearly impacts how it is understood.

Consumers Go For "75 % Lean", Not For the Twin "25 % Fat"

On a lighter note, people's emotional reaction to the fat content of meat varies, depending on whether it is presented as *"75 % lean"* or *"25 % fat."*

In a study[9] conducted by psychologists Irwin P. Levin and Gary J. Gaeth, from the University of Iowa, it was demonstrated that consumers reacted more favourably towards beef labelled *"75 % lean"* than the same beef labelled *"25 % fat"*.

When consumers had actually tasted the meat, the magnitude of this information-framing effect lessened.

---

[9] Levin, I. P. and Gaeth, G. J., 1988

More On Frames and Framing

 A Small Dose of Theory On Frames

Along with many scholars, George Lakoff[10], Professor of Cognitive Science and Linguistics at the University of California at Berkeley, stresses that we conceive, mostly unconsciously, in terms of systems of structures called *"frames"*. Each frame is a neural circuit that is physically located within the brain. We use our frame circuitry systems to understand everything, and we reason using logic that is accepted within the context of these frames.

 A Helpful Tip On Communication Framing

When you communicate on prevention, consider stressing losses (if prevention is ignored) rather than gains (if prevention is implemented), building on the human bias of loss aversion.

Let's recall the number of occasions on which health authorities and doctors try to convey the right message or convince people to undertake a new treatment, or pill, or surgery.

Writing and Speaking

As claimed by Steven Pinker, language is an instinct[11] and our brain hosts circuits specialized for speech and for assembling words according to grammatical rules, to obtain spoken sentences.

Robin Dunbar analysed human conversations and concluded that 60% of our time was spent on gossip about relationships and other people's personal lives: *"It is suggested that language evolved to allow individuals to learn about the behavioural characteristics of other group members more rapidly than is possible by direct observation alone."*[12] No surprise that gossiping and reading magazines about the rich and famous still constitutes a large share of human activities.

---

[10] Lakoff, G. and Johnson, M., 1999; and "How we Talk About the Environment has Everything to do with Whether We'll Save it" retrieved from http://www.tikkun.org/article.php/2009052014051976 on December 12th, 2012
[11] Pinker, S., 1997
[12] Dunbar, R. I. M., 1993

By contrast, writing appeared much later in the story of mankind, in Mesopotamia, about six thousand years ago. It is most unlikely that evolution would have had enough time to sculpt refined neural modules dedicated to writing, in our brains and minds.

According to Stanislas Dehaene, Professor in Experimental Cognitive Psychology at the Collège de France, pre-existing brain circuits had to be recycled to handle writing[13]. This means that our circuits dedicated to writing will not be as sophisticated in the way they accomplish their function as those for speaking.

 Useful Tips to Write a Message

Speaking and listening are more natural operations than writing and reading.

It is no surprise that writers recommend the following tricks: use dialogues, favour musicality in texts (such as alliterations), read your text aloud. Advertisers often exploit the trick of conveying a claim in the form of a short and melodious song or jingle. Does a famous chewing gum ad ring a bell?

Learning

 A Small Dose of Theory On Learning

Michael Gazzaniga stresses that:

When we think we are learning something, we are only discovering what already has been built in to our brains.[14]

Therefore, some kinds of learning may be hard when selection has not had sufficient time, nor adaptive reason, to shape the corresponding neural circuits, as exemplified by reading or higher mathematics, that are difficult to learn.

This simple idea implies that there ought to be impinging on us information and circumstances that are difficult to learn very well.
Michael S. Gazzaniga[15]

---

[13] Dehaene, S., 2007
[14] Gazzaniga, M. S., 1992
[15] Gazzaniga, M. S., 1992

Such an assertion echoes Steven Pinker's claim that is hard to mobilize a mental module to handle situations it has not been conceived for[16].

Our aptitude for simple arithmetic seems to be innate. As recalled by Pinker, psychologists have shown that babies have the capacity to distinguish two to three objects when they are only one-week old, or to do simple calculations when they are five months old.[17]

Encourage Organ Donation By Making It the Default Option (Nudges)

Recent trends in public decision-making stress the importance of matching the message with the characteristics of the biased mind rather than trying to educate or edify people with campaigns intended to reach them on another level.

Gerd Gigerenzer describes a *nudge*, a trick based upon behavioral science, namely the status quo bias, to convince people to donate their organs to health authorities[18]. Every year, an estimated 5,000 Americans die waiting for an organ donor who never materialises. As a consequence, a black market in kidneys and other organs has emerged as an illegal alternative. In France, the situation for people awaiting donor organs is much better, and many more gravely ill people receive the new organs they need.

So why are only 28 % of Americans potential organ donors in comparison with a striking 99.9 % of the French? Do the French have a higher moral consciousness, or are Americans perhaps less well informed about the shortage? The answer cannot be found by examining differences in national personality traits or access to information. Rather, the majority of Americans and French seem to employ the same default mental shortcut, which states, "If there is a default, do nothing about it"—a clear example of the status quo bias.

In the United States, the legal default is that nobody is a donor without registering to be one. You need to opt in and make a conscious decision about what is to be done with your organs in the event of your untimely demise. Who enjoys thinking about the possibility of something bad happening to them while they are still young and healthy enough for their organs to be suitable for transplant? As we saw, most people assume that they are less at risk of accident and more likely to live to a ripe old age than the average. In France, the situation with respect to organ donation is very different. *Everyone* is a potential donor unless they opt out of the default mode and make the conscious decision that their organs are not for sharing. As this also involves thinking about their untimely demise, it is not surprising that most people stick with the default.

---

[16] Pinker, S., 1997
[17] Pinker, S., 1997
[18] Thaler, R. and Sunstein, C. R., 2008

In this example, behavior is a consequence of the default mental shortcut and of the legal environment, leading to the striking contrast between the two countries. The implication for policy-making in this case seems to be quite clear. By making organ donation the default situation, thousands of people can be taken off waiting lists; thousands of lives can be saved and thousands of hearts can continue to beat.

The problem of the perennial shortage of donor organs shows how important it is to play this right. If authorization is the default decision (somebody who would not want to donate has to fill out a special request saying NO), then donations are much more numerous than when everyone has to authorize donations explicitly. Based upon this example and others, Gigerenzer claims that:

> *The donor problem illustrates this conjecture, where thousands of lives could be saved every year if governments introduced proper defaults rather than continuing to bet on the wrong "internal" psychology and send letters to their citizens.*

This is an insightful statement showing that pedagogy, in its common-belief ambition to enlighten us by directly addressing our noble innate feelings, can sometimes be less efficient than changing the environment in which the message is sent.

Handling Probabilities In Communication
Outcomes Count More Than Probabilities

When you think of a problem regarding nuclear energy, you visualize a possible accident, but pay little attention to the corresponding probability. When you buy a lottery ticket, you see the potential jackpot without considering the infinitesimal probability of success. In lotteries as in risky technologies, possible outcomes call more to mind than the probabilities that they materialize. Jeryl L. Mumpower, from Texas A&M University, stresses strong structural similarities between risks from technological hazards and big-purse state lottery games[19].

Risks from technological hazards and state lotteries can equally be described as low-probability, high-consequence events. In the case of lotteries these are high-consequence positive events, vividly portrayed in advertisements showing happy winners in private jet planes or on remote tropical islands. According to Mumpower, typical communications about state lotteries could be a

---

[19] Mumpower, J. L., 1988

useful source of inspiration for opponents of risky technologies. By contrast, defendants of risky technologies should consider that arguing on probability will have little effect.

Reframing the Classic Harvard Medical School Test

Here are ways to reframe specialized health messages into edible pieces. Take the famous brainteaser called the medical diagnosis problem, as it was initially conducted in 1978 at the Harvard Medical School among 20 house officers, 20 fourth-year medical students, and 20 attending physicians, selected in 67 consecutive hallway encounters at four Harvard Medical School teaching hospitals, by cardiologist S. Ward Casscells, Arno Schoenberger, and Thomas B. Graboys.[20]

We suggest that you try to answer yourself:

 Test: Do Better Than Harvard Students and Staff

If a test to detect a disease whose prevalence is 1/1000 has a false positive rate of 5 %, what is the chance that a person found to have a positive result actually has the disease, assuming that you know nothing about the person's symptoms or signs?

Only 18 % of Harvard medical school students and staff answered "2 %", the correct answer, while an astounding 45 % of them answered "95 %".

Now, Leda Cosmides and John Tooby[21] presented the same problem with frequency formats instead of probabilities:

- *1 out of every 1000 Americans has disease X.*
- *A test has been developed to detect when a person has disease X.*
- *Every time the test is given to a person who has the disease, the test comes out positive.*[22]
- *But sometimes the test also comes out positive when it is given to a person who is completely healthy. Specifically, out of every 1000 people who are perfectly healthy, 50 of them test positive for the disease.*[23]

---

[20] Casscells, W., Schoenberger, A. and Graboys, T. B., 1978
[21] Cosmides, L. and Tooby, J., 1996
[22] i.e., the "true positive" rate is 100 %
[23] i.e., the "false positive" rate is 5 %

Now, with this formulation, 56% gave the correct answer.

With the former framing, there are many more correct answers. But we can do even better! Elke Kurz-Milcke, Gerd Gigerenzer, and Laura Martignon[24] report various examples of how a visual description of medical statistical issues—with individuals presented as squares, circles, or small figurines—improves understanding and thereby also improves the answers to tests.

Picture: how to solve the Harvard test more easily

Let's comment on the illustration above. We display 25 rows and 40 columns of dots, hence 1000 points representing 1000 individuals. The red point (bottom left) corresponds to the one out of a thousand Americans having the disease. In this way, we express visually that the prevalence of the disease is 1/1000.

The blue shaded area to the left displays the second and third columns with 25 rows. It thus contains 50 points, representing the fifty individuals within the sample who will give a positive result in the test, even though they do not have the disease. This represents the so-called "5% false positive rate".

We can visualize the 51 individuals being positive at the test, either rightly (1 red), or wrongly (50 blue).

Therefore, the probability of having the disease if you give a positive result in the test, is 1/51. So the correct answer should be 100/51 (in %), which is slightly less than 2%

That's all folks!

---

[24] Kurz-Milcke, E. K., Gigerenzer, G. and Martignon, L., 2008

Deciphering Questionnaires

From time to time, we are confronted with questionnaires: renewing a personal loan or applying for a new insurance policy, answering an opinion survey, etc. In the search for happiness, college students answered a survey that included two questions:

- How happy are you with your life in general?
- How many dates did you have last month?

Initially, there was very little correlation between the two questions. However, when the order of the questions was reversed, everything changed! When the dates were asked about first, the students' evaluation of how happy they were became charged with emotion, providing them with a more precise, more vivid, and less abstract example of happiness.[25] The context of the questionnaire (at work, on vacation), the order in which questions are given (general and neutral first, before becoming more personal), the framing of the question itself (the wording)…all these parameters influence the way we answer. The very theme of the questionnaire may initiate some kind of personal defence strategy.

 Helpful Tips On How to Decipher Answers to Questionnaires

When answering a bank or sociological questionnaire, or when interpreting opinion surveys, be cautious of the power of framing. In a disturbing French questionnaire by Jeannine Richard-Zapella[26], Professor of Literature at University of Picardie-Jules Verne, the following equivalent questions, addressed in 1969, gave significantly different answers:
81 % of the people answered "yes" to: *"Croyez-vous en Dieu?"*
(A direct way to ask: *'Do you believe in God?'*)
66 % answered "yes" to: *"Est-ce-que vous croyez en Dieu?"*
(*'Is it that you believe in God?'* A common way to ask the same thing in French, but less direct)
Could it be that the phrase "Is it that" introduces a certain distance with respect to the core of the question?

Anonymous But Still Hyper Social

In *La Recherche du Temps Perdu*, French writer Marcel Proust wrote that our social personalities are created by what others think. This is a good way to ex-

---
[25] Frederick, S.W., 2002
[26] Richard-Zappella, J., 1990

press people's tendency to present themselves favourably according to current cultural norms. This seems to be the case in many habit surveys.

Delroy L. Paulhus, from the University of British Columbia "defines response biases as any systematic tendency to answer questionnaire items on some basis that interferes with accurate self-reports."[27]

For example, when confronted with the question, *"Do you use drugs/illicit substances?"* the person asked may be influenced by the fact that controlled substances, including the more commonly used marijuana, are generally illegal, and indeed looked down upon by a certain part of the population. As a result, they may feel prompted to answer that they do not use drugs at all, or at least to play down the frequency of their use of such a drug: *"I only smoke marijuana when my friends are around."*

Similarly, surveys of adult sexual habits tend to show men engaging in many more sexual encounters than women. In most societies, women are penalised to varying degrees for promiscuous sexual activity, whereas men are not (often, in fact, the reverse), creating a situation in which members of both sexes are liable to *lie,* even to themselves, about their sexual behavior, in order to conform to social norms.

A common way to assess this bias is by submitting a questionnaire that mixes innocuous questions: "Do you speak more than one language reasonably well?", and sensitive ones like: "Have you ever, even once, taken something from a store without paying for it?"[28]

Paulhus[29] proposed just such a test that could be used to evaluate what he calls *"socially desirable responding".* By answering the *Balanced Inventory of Desirable Responding (BIDR),* each individual is determined either to be "self deceptive" or "making a favourable social impression" by cross-analysing more than 40 questions of the kind: *"I have done things that I don't tell other people about",* or *"I have sometimes doubted my ability as a lover",* or even *"I never read sexy books or magazines".*

Such a bias towards social desirability or political correctness can be so entrenched in our mind that traces of it have been found even when a questionnaire has been addressed anonymously. One might guess that, if people sometimes make a false statement, that can be put down to the weight of social correctness, whence anonymity ought to dramatically reduce concern overself-representation. And yes, when questionnaires and surveys are addressed anonymously, we tend to be less preoccupied by what others might think. Our habits are no longer being monitored. Indeed, anonymous questionnaires usually gather more results

---

[27] Paulhus, D. L., 1991
[28] Paulhus, D. L., 1991
[29] Paulhus, D. L., 1991

about socially inappropriate habits[30]. An intriguing study made by Anthony D. Ong, from the University of Southern California, and David J. Weiss, from California State University, Los Angeles, involving 155 undergraduate students, focused on cheating, a stigmatized although fairly common habit among students. The study demonstrated that, under anonymous conditions (no name given), more students admitted they cheated on their partners than under less strict confidential conditions (name given to the experimenter but not traced[31]). More intriguing, Ong and Weiss also found that, even under anonymous settings, 26% of known cheaters did not admit that they had cheated.

Reframing an Abstract Problem In Terms of Social Interactions

Wason Cards

A series of experiments initiated by British psychologist Peter Wason at the end of the sixties[32]—the so-called "Wason task" regarding our logical abilities—is the starting point for a discussion about the fascinating role of context in our comprehension.

Suppose I show you four cards on the table: E, 7, K, and 2.
I tell you that the cards have letters on one side and numbers on the other. Wason asked the following:

Which cards do you necessarily have to turn around in order to check whether the following rule holds for the set of 4 cards:
"If one side of a card exhibits a vowel, its other side must exhibit an odd number?"

Most subjects (>85%) give a wrong answer, generally pointing to cards E and 7, or only to card E.[33]

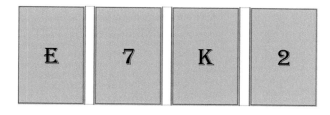

Picture of Wason card puzzle

---

[30] Ong, A. D. and Weiss, D. J., 2000
[31] Confidentiality guarantees that no traceable record of the participant's data will be disclosed. Ong and Weiss knew who had really cheated or not!
[32] Wason, P. C., 1968
[33] The correct answer is to turn around card 2 in addition to card E

Several psychological experiments based on this experiment revealed that the problem was more easily solved once it was reformulated in a social context. For instance, Laura Martignon and Stephen Krauss relate a study[34] in which *"the abstract context of the binary features "vowels vs. consonants" and "even vs. odd numbers" is replaced by a context typical of social contract situations".*[35]

Consider this new formulation. Now each card represents an envelope to be sent by mail: the destination is written on one side (vowel) and the value of the stamp stands on the other side (number).

The rule is now: "If the letter goes to Europe, then it requires at least $ 2.50". Most subjects give the right answer.

Martignon and Krauss relate this observation to the claim from Cosmides and Tooby, that the mind is equipped with a *"cheating detection module"*[36].

In a study which involved determining whether or not cheats could be detected, research subjects were best able to figure out when someone was cheating when the issue was framed as a social contract. On the website of the Center for Evolutionary Psychology at the University of California in Santa Barbara, we found the following story, turning the Wason card problem into a social contract rather than an abstract issue. This specific contract has to do with teenagers following a simple rule given by their parents:

> *Teenagers who don't have their own cars usually end up borrowing their parents' cars. In return for the privilege of borrowing the car, the Goldsteins have given their kids the rule, "If you borrow my car, then you have to fill up the tank with gas."*

Then you are presented with four cards. One side says the kid borrowed the Goldsteins' car and the other side mentions whether the teenager filled up the tank with gas on the particular day when he borrowed his parents' car.

> *Which of the following cards would you definitely need to turn over to see if any of these teenagers are breaking their parents' rule?*
> *"If you borrow my car, then you have to fill up the tank with gas."*
> *Don't turn over any more cards than are absolutely necessary.*

---

[34] Johnson-Laird, P. N., Legrenzi, P. and Legrenzi, M. S, 1972. Reasoning and a sense of reality. British Journal of Psychology, 63, 395–400
[35] Martignon, L. and Krauss, S., 2009
[36] Cosmides, L. and Tooby, J., 1992

| borrowed car | did not borrow car | filled up tank with gas | did not fill up tank with gas |

This example illustrates that a solution is easier to determine when shown in the context of a social contract rather than as an abstract text. The answer is that you need to turn over cards one and four, not one and two, as people often choose when the Wason task is presented with abstract notions on the cards.

Steven Pinker proposes another social presentation in which you are supposed to be a server in a bar, enforcing the rule: *"If a person is drinking beer, he must be eighteen or older."* Which are the two questions you should ask to make sure the rule is respected? You will have to know the drink somebody is having and her age. *"Which do you have to check: a beer drinker, a Coke drinker, a twenty-five-year-old, a sixteen-year-old? Most people correctly select the beer drinker and the sixteen-year-old."*

It Pays to Frame Abstract Matters As Vivid Metaphors

Eight friends go out for dinner and before eating they decide how they are going to split the bill. Will everyone pay for what they eat, or will they simply divide the cheque by eight? Richard Thaler compares the management of the US federal budget to the check after a restaurant dinner[37]. The US federal budget funds different ministries and divisions and, unfortunately, the surest way to get a bigger budget is to split the overall budget between the ministries. Nobody feels responsible as nobody has to justify their own expenses.

Understanding how decisions are made in abstract and complex issues such as public spending is much easier when the problem is framed in a more mundane context. Going to the restaurant and paying the check appears to be a powerful metaphor, triggering the right chord: our bodily need for food.

How Our Body Insinuates Itself Into Our Mind

Embodied Language and Thinking

The body, and especially motor movements, can have a big impact on the mental shortcuts we default to. Research has demonstrated that people are

---
[37] Thaler, R., 2000

more likely to accept propositions when they are nodding their heads up and down than when shaking them from side to side.[38] The mind drives the body, but the other way around, too.

Picture: a hat, a hand, two legs, and the biased mind

What Makes a Domestic Appliance User-Friendly?

In terms of understanding visual instructions in general, Sarah Davies, Helen Haines, Beverley Norris, and John R. Wilson, all from the University of Nottingham, have found that the ones that people find most difficult to understand are the most abstract in nature, such as the tumble-drier and the "on" switch[39].

Around the world, countless living rooms contain countless TV remote controls of which only three or four buttons are ever used, because the owners have no idea what the rest of them are actually for.

To make it easier to understand, the manufacturer has made a point of drawing a plan at the bottom of the hob which reproduces the spatial distribution of the 3 hobs.

---

[38] Wells, G. L. and Petty, R., 1980
[39] Davies, S., Haines, H., Norris, B. and Wilson, J. R., 1998

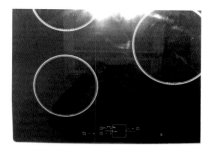

Picture: an induction hob (seen from above)

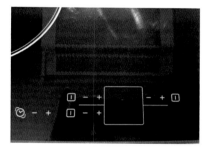

Picture: detail showing a diagram of the 3 hobs

Designers of TV and kitchen hobs should try to get through to the body in our mind.

Our Understanding Is Grounded In Body-Based Metaphors

George Lakoff and Mark Johnson, Professor of Liberal Arts and Sciences at the University of Oregon, developed the thesis according to which our understanding is grounded in metaphors, the stuff of which is our bodily experience[40]. Time is conceived as movements in space in such expressions as "it took a *long* time", *"before"* and *"after"* the event; ideas are *"containers"* that can be *"grasped, packaged, sent"* and *"transmitted"*, etc.

Our mind hosts a specific area devoted to action verbs[41]. Now, it is no coincidence that manuals on writing—like the instructive *Style: Lessons in Clarity and Grace,* by Joseph M. Williams, late Professor of English Language and

---

[40] Lakoff, G. and Johnson, M., 1980
[41] Willems, R. M., Hagoort, P. and Casasanto, D., 2010

Literature at the University of Chicago—recommend, for clarity, the use of verbs to describe important actions.

On the *Copyblogger* website[42], we find advice by Gregory Ciotti to use action verbs rather than adjectives:

> *Verbs get specific and are harder to ignore, especially in a vain world where everybody describes themselves with the same trite adjectives.*

Mrs Dee Leopold, managing director of M.B.A. admissions at Harvard Business School, describes how Harvard makes admissions decisions:

> *The best recommendations have a lot of verbs. They say, "She did this," versus adjectives that simply describe you.*

When someone reads "grasping the idea", zones corresponding to the physical action of grasping are activated in the premotor cortex of the left hemisphere[43].

 A Helpful Tip On Using Action Verbs and Metaphorical Words

Instead of *"I understand"*, use *"I have grasped the concept"*, a more vivid expression for our minds

Instead of
"We analyzed the figures and found interesting facts"
say
"We explored the figures, crunched out numbers, and dug out the most valuable gems for you."

In spoken language people use metaphors to catch the attention.

Instead of "he is lazy", say "he is bone-idle".
Instead of "they tried to bribe someone", say "they tried to grease someone's palm",

---

[42] http://www.copyblogger.com/scientific-copywriting/7 Scientifically-Backed Copywriting Tips, by Gregory Ciotti
[43] Aziz-Zadeh, L., Wilson, S. M., Rizzolatti, G. and Iacoboni, M., 2006

And instead of "she made a mistake", try saying "she was barking up the wrong tree".

Interestingly, it seems that our mind transforms a passive sentence into an active form when memories are stored. Michael Gazzaniga reports[44] experiments by psychologists John Anderson, Professor of Psychology and Computer Science at Carnegie Mellon University, and Gordon Bower, Emeritus Professor at Stanford University. Subjects were read the statement *"The girl was kissed by the boy"* and asked, 2 min later, which of four sentences were consistent with it. Subjects reacted more rapidly to the active form *"The boy kissed the girl"* than to the original passive form.

You're Right to Talk to the Left Ear

What we see with our right eye is sent to the left brain, and conversely. Our left hemisphere is better at handling texts and symbols than the right hemisphere, specialized in images. This brain asymmetry has consequences on how we communicate with others. Among many other intriguing examples, Evan W. Carr, Sebastian Korb, Paula Niedenthal, and Piotr Winkielman demonstrated that people not only produce spontaneous expressions earlier on the left side of the face, but they also perceive the expressions of others as more spontaneous when they start on the left side of the face[45].

Here are two tips, one light-hearted, one more serious.

> **A Helpful Tip On Making Slides For Presentations**
>
> What we see with our right eye is sent to the left brain, and conversely. The left hemisphere is better at handling texts and symbols than the right hemisphere, specialized in images. Therefore when you make a visual presentation with slides, we advise you to position images to the left of your document and comments to the right.

---

[44] Gazzaniga, M. S., 1992
[45] Carr, E. W., Korb, S., Niedenthal, P. M. and Winkielman, P., 2014

 Somebody is watching you

 Somebody is watching you

Pictures: which is the most striking?

 A Helpful Tip For Lovers

When you approach your beloved, better to whisper words of love in the left ear. Indeed an astonishing study has revealed an advantage for words of emotion spoken in the left ear.[46]

The Watchful Eye Bias: "l'oeil était dans la tombe et regardait Caïn" (Victor Hugo)

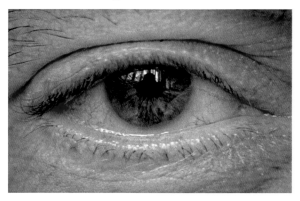

Picture: I am watching you

---

[46] Sim, T. C. and Martinez, C., 2005

In a poem entitled *"La Conscience"*, French writer Victor Hugo well captures the compelling power of the eye, pursuing the remorseful Cain ("The eye was in the grave and looked at Cain").

Costas Panagopoulos, Associate Professor of Political Science at Fordham University, gives several intriguing examples of how the visual representation of an eye affects our behavior: *"posters displaying images of eyes caused people to be more likely to remove litter from cafeteria tables"*, and *"pictures of eyes placed next to an "honesty box" in a university psychology department coffee room tripled employee donations"* [47].

As we stressed earlier, and as can be seen for example in questionnaires aimed at assessing sensitive human habits, we are *hyper social* animals. Evolution has shaped mechanisms that monitor and preserve interactions in groups.

Quick Interpretations of Faces

We see patterns everywhere. Especially faces and eyes;-)

Picture: an inspiring sky in French Normandy

According to Michael Gazzaniga, *"The modern human is a bundle of special-purpose systems that allow us to communicate, evaluate facial expressions, make inferences, and interpret feelings, moods, behaviors, and all the rest."* [48]

It is a highly adaptive advantage to be able to interpret feelings quickly and accurately from facial expressions, or intentions and moods from voice tones.

---

[47] Panagopoulos, C., 2014
[48] Gazzaniga, M. S., 1992

According to Darwin: *the movements of expressions (...) reveal the thoughts and intentions of others more truly than words, for they are more difficult to falsify.*

Picture: plate depicting emotions of grief from Charles Darwin's book
The Expression of the Emotions in Man and Animals

When Darwin was developing what would become his famous theory of evolution[49] —which asserts that humans descend from other animals and share many attitudes and behaviors with them—he did the groundwork to demonstrate that a lot of our facial expressions can be traced to our primate predecessors: feelings of fear, joy, terror, and lust[50]. We still show our teeth when we are angry, perhaps because our animal predecessors did so to indicate that they might be about to bite during an aggressive encounter. Darwin was a pioneer in demonstrating the adaptive nature of human beings.

In the same vein, trivial introductory sentences like the ones listed below do not convey their true meaning from their wording, but rather from the tone of voice. One can indeed infer a great deal of information about feelings, moods, and intentions from the tone of voice. Marketers advise salespeople to smile when they phone potential clients, precisely because it has a favorable impact on the tone of voice. Women's voices rise in pitch during the ovulation phase.[51]

Although introductory expressions and polite remarks may differ slightly from one language to another and for cultural reasons (see below), they share the fact that their wording alone usually conveys little meaning. Perhaps they are just a quick and dirty way of interpreting moods and intentions without the labor of a full interrogation or a long observation.

---

[49] Darwin, C., 1859.
[50] Darwin, C., 1998 (orig. 1872)
[51] Buss, D., 2014

In casual Spanish from Argentina:
Hola, che que tal, como andas? (Hey man, what's up?)
Barbaro y vos? (Great and you?)
Bien, gracias (Fine, thank you)
In Swahili
Jambo, habariya (Hello, good morning, how are you)
asubuhi, unaendeleavipi? (Morning, thank you, and you?)
Nsuri, nayako (I'm OK, thanks)
In French
Bonjour, comment allez-vous? (Hello, how are you?)
Bien, merci, et vous? (Fine, thank you, and you?)
Bien, merci. (Doing fine, thanks)
In Chinese
你好。*nihao* (Hello)
很高兴见到你! *hengaoxing jiandaoni!* (Very nice to meet you)

Herman Chernoff's Precursors of Smileys

Long before *smileys* were thought of, the famous statistician Herman Chernoff came up with human faces to express data[52]. Because people are naturally good at seeing faces and noticing small changes in them, Chernoff suggested presenting data in the shape of human faces. Features such as the nose, eyes, and mouth can represent variables, communicating information by where they are placed. He noted that some features, or variables, are given more importance, such as the size of the eyes and the position of the eyebrows.

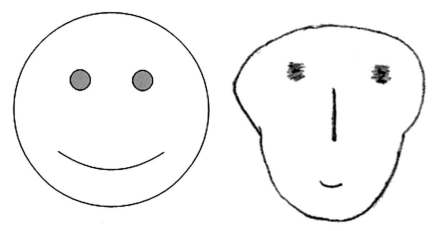

Picture: a modern smiley (on the left) and a Chernoff-like face (on the right), inspired from the 1973 Chernoff faces.

---
[52] Chernoff, H., 1973

9
Epilogue: Does It Really Pay to Weigh Up Our Biases?

Now we know a little more about how the human mind is biased. We all love anecdotes and images and we played on this by giving in to our natural propensity to turn the scientific material of *The Biased Mind* into palatable and edible pieces. Did we succeed in making the reading pleasant and enriching? Will you now communicate more efficiently and make better informed decisions?

We really hope so. ☺ Indeed, we, the authors, were not that lucky. ☹

We initially planned to finish *The Biased Mind* within 6 months.

That was 7 years ago!

Detailed Contents

1 Introduction 1

2 Embarking On the Mind Tour 5
Who's the Boss? 6
    Making Virtuous Choices 6
    The Dumbledore Pact 7
    Hero But Shy With the Ladies? 8
    When Dr Jekyll Becomes Mr Hyde 9
    The Multi-Modular Mind Hypothesis 10
Please Alleviate My Cognitive Burden 11
    The Magical Number 7 11
    Happiness Is a Matter of (Not Too Much) Choice ... 12
    The Social Number 150? 13
The Mind As a Survival Kit 15
    The Mind As an Adaptive Toolbox 15
    Our Biases Reflect Human Ecological Rationality ... 16

3 Better Be Paranoid to Survive 19
It Was Scary In Flintstone... 20
    The World Is Populated With Survivor's Heirs ... 21
    Our Hunter-Gatherer Parents Thrived In the Savannah ... 21
    SSSSSS...Sinuous Snakes Still Scare! 24
...It's Scary Now! 27
    What Makes a Landscape Friendly? 27
    Between the Rock and a Hard Place 29
    Proverbs Convey Cultural Risks 31
    Paris In the Summer When It Sizzles 33
    Needles Can Breach the Body's Defences 33
    We Are Scared of Transgression 35
    We Have Not Been Hard-Wired to Cope With Modern Risks ... 36

Unknown New Risks Surround Us	37
Unknown Unknowns Are Difficult to Pinpoint	37
Are We Paranoid Optimists?	38
Sometimes It Is Best to Be a Little Paranoid! (Paranoia Saved Our Optimistic Parents)	39
Louder Is Closer	40
Descent Illusion	41
Our Radar Minds Are Tuned Like a Smoke Detector	42
Asymmetries Induce Strong Biases	43
The Life-Dinner Principle (the Predator Runs for Dinner, the Prey for Life)	44
Seeing Storms Behind the Clouds	45
The Coolidge Effect	46
Romance, Sex, and Other Biases	47
The Blackstone Ratio In Courtrooms	47

4 We Like Things the Way They Are ... 49

Losses Loom Larger Than Gains	50
Paying One's Taxes At Source Feels Less Painful	52
Don't Wrap Your Christmas Gifts In a Single Package	52
Mental Accounting	53
We Are Biased Towards the Status Quo	54
Suppose You Are Compelled to Play Russian Roulette	54
Hands Off My Stuff!	55
Coca Cola Fans Leaned for the Old Coke Classic	56
A Small Town In Germany	57
The Drubeck Brothers' Story	57
"Tell Me Something I Didn't Learn In Hotel School"	58
It Took 400 Years Before We Used Lemon Juice to Avoid Scurvy	59
When We Anchor Our Assessment On Mere Fortune	59
Did Gandhi Live More Than 140 Years?	60
When Less Is More	60
We Succumb Over and Over Again to Committing Ourselves	61
Desperately Seeking...Confirmation	62
When Karl Popper Caught Psychiatrist Alfred Adler In Confirmatory Bias	63
How to Influence the Perceptions of Fidel Castro	64
Love Is Blind, But the Neighbours Aren't	64

5 Our Detective Mind Grasps Clues and Narrates 65
Cherry-Picking and Connecting the Dots 66
    The Barnum Effect ("We've Got Something for Everyone") 66
    Cold Readers Can Mystify Strangers by Knowing All About Them!...... 67
    How Venusian Artists "Cold Read" Female Targets 68
    Heuristics and Mental Shortcuts Fill the Dots 69
    Mental Shortcuts Ground Optical Illusions 69
    "Watch the Borders!" Wrote Former FBI Director Edgar J. Hoover...... 71
    Cooperative Efforts In Conversation 71
    What Makes a Good Alibi?... 72
What If?... 74
    What If...? If Only I Had... 74
    When Bronze Feels Better Than Silver............................... 75
    The Charismatic Spanish Bullfighter Yiyo 76
    Hindsight Is 20/20 76
    You'd Never Forgive Yourself 77
Monkey See, Monkey Do... 78
    Social Proof 78
    Milgram's Experiment: On 42nd Street, Look Up At the Sky
    and Everyone Will Follow Suit!.................................... 78
    Would "Do As Others Do" Stand At the Root of the Evolution
    of Cooperation?... 79
I Am Not Superstitious, That Brings Bad Luck 81
    We're All Control Freaks.. 82
    Bad Things Don't Happen to People Like Us........................ 82
    Magical Thinking... 83
    Homeopathy: The Magic of Almost Nothing........................ 84
    Does God Answer Your Prayers? 84
    Was the German Bombing of South London Done At Random?........ 86
    The World Did Not End In 2012, But the Mexican Tourism
    Industry Flourished!.. 86

6 Images Call More to Mind Than Words and Numbers................. 89
Thinking With Our Guts.. 90
    On November 16, 1532, In Cajamarca, the *Conquistadores* Faced
    Unknown Fear 90
    The Fast Track to Fear 91
    The Slow Road to Thinking, Reasoning, and Consciousness........... 93

Fear of Flying	93
What Is the Riskiest Part of a Plane Trip? Driving to the Airport!	95
Insurance Feelings	96
When Blackheads Matter More Than Cancer	97
We Do Not See a House But a Handsome Or an Ugly house	97
Chickening Out	99
Our Mental States Blow Hot and Cold	99
Walkin' On Sunshine	100
Customers Leave Larger Tips On Sunny Days	101
Soccer Results Impact Wall Street	101
The Peak-End Rule: Peak and End Experiences Matter More Than Duration	101
"I Know a Brazilian Man Who…", Or the Power of Anecdotes and Vivid Testimonies	102
When Two First Ladies Prompt More Cancer Detection Than Dry Statistics	103
The Availability Heuristic	104
Which City Is the Biggest? San Diego Or San Antonio?	105
The French Eat Snails, Not Slugs	105
In the Mind's Eye	108
Jack Was Killed By a Semi-trailer	108
Images, Words, and Emotions	109
When Detroit Looms Larger Than Michigan	111
The Larger Bowl Looked More Inviting (Though It Offered Fewer Prospects)	111
Media Coverage of Epidemics Calls Images to Mind	112
Ebola Virus Imagery Held Off the 1995 Epidemics In a Science Fiction Limbo	113
Lies, Damn Lies, and Statistics	114
Tom Is Tall Because He Is Heavy	114
Adding a Small Loss Makes the Bet More Attractive!	114
Distinction Bias	115
The Gambler's Fallacy (Gamblers Yearn For the Good Outcome That Will Offset All the Bad Ones)	116
The Base Rate Fallacy	117
Pill Scare	118
Are Enemies At the Door?	118
23 People and 2 Birthdays	119
When 20 Out of 100 Is Not Equal to 20 %	119

	Rebukes Would Boost Learning, While Praise Backfires... Really?.	120
	Feelings and Statistics Are Poor Bedfellows	121
	Intuition Did Not Break the Sound Barrier	121
	The Minimum Time Trajectory Defies Intuition.	122

7 How to Balance Pros and Cons, and Other Helpful Hints 123

Five Anecdotes On How Decisions Are Reached 123

    J. C. Penney Hired Executives "On Salt". 123

    US President Ronald Reagan Tested Visitors with Jelly Beans 123

    Economics Nobel Prize Markowitz Did Not Apply His Theory
for His Own Investments. 124

    C'mon, This Is Serious!. 124

    Garry Trudeau's Lists 124

Piece-of-Cake Algebra Predicts Better Than Personal Judgment 125

    How "Unromantically" Did Charles Darwin Decide to Marry 125

    Benjamin Franklin Advocates "Moral Algebra"
to Weigh Up Pros and Cons 127

    Paul Meehl's Review of Evidence 127

    Dry Linear Models Could Usefully Complement Smart
Expert Categorization. 128

    Marks Do Better Than Interviews (Though Both Are Weak
Predictors In the Absolute). 129

    An Aside On Assessments and Admissions. 130

    Crude Arithmetic Can Predict Marital Happiness 132

Experts Excel In Extracting Relevant Features and Categorizing 133

    Intuitive London Magistrates 134

    Beyond Words, Do Experts Agree On the Occurrence of Unlikely
Earthquakes?. 134

    Expert Burglars Rely On a Few Selective Cues. 135

    Become a Wine Tasting Expert 136

Overconfident Experts 138

    Even Cold-Blooded Experts Betray Over-Optimism. 138

    Watch Out! Better Not Read This Chilling Study Before Embarking
On an Airplane 139

    The Monty Hall Problem Baffled Mathematical Experts. 140

    The Planning Fallacy, Or How We Deceive Ourselves 141

8 I Frame, You're Framed 143

Why Don't We Stop At a Green Light?. 145

Emphasizing Losses, Gains, and Anchors 149

    Changing Two Words Makes a Striking Difference. 149

Consumers Go For "75 % Lean", Not For the Twin "25 % Fat"	150
More On Frames and Framing	151
Writing and Speaking	151
Learning	152
Encourage Organ Donation By Making It the Default Option (Nudges)	153
Handling Probabilities In Communication	154
Outcomes Count More Than Probabilities	154
Reframing the Classic Harvard Medical School Test	155
Deciphering Questionnaires	157
Anonymous But Still Hyper Social	157
Reframing an Abstract Problem In Terms of Social Interactions	159
Wason Cards	159
It Pays to Frame Abstract Matters As Vivid Metaphors	161
How Our Body Insinuates Itself Into Our Mind	161
Embodied Language and Thinking	161
What Makes a Domestic Appliance User-Friendly?	162
Our Understanding Is Grounded In Body-Based Metaphors	163
You're Right to Talk to the Left Ear	165
The Watchful Eye Bias: "l'oeil était dans la tombe et regardait Caïn" (Victor Hugo)	166
Quick Interpretations of Faces	167
Herman Chernoff's Precursors of Smileys	169
9 Epilogue: Does It Really Pay to Weigh Up Our Biases?	**171**
Detailed Contents	**173**
Bibliography	**179**

Bibliography

Appleton, J., The Experience of Landscape. John Wiley and Sons, 1975.
Ariely, D., Predictably Irrational: The Hidden Forces That Shape Our Decisions. Harper Collins, 2008.
Axelrod, R. and **Hamilton**, W. D., The Evolution of Cooperation. Science, 211: 1390–96, 1981.
Aziz-Zadeh, L., **Wilson**, S. M., **Rizzolatti**, G. and **Iacoboni**, M., Congruent embodied representations for visually presented actions and linguistic phrases describing actions. Curr Biol., 16 (18):1818–23, 2006.
Balling, J. D. and **Falk**, J. H., Development of visual preferences for natural environments. Environment and Behavior, vol. 14, no 1, 5–28, 1982.
Bar-Hillel, M. and **Ben-Shakhar**, G., The A Priori Case against Graphology: Methodological and Conceptual Issues. In Connolly, T., Arkes, H. R. and Hammond, K. R. (Editors), 2000.
Barg, F. K. and **Grier**, S. A., Enhancing breast cancer communications: A cross-cultural approach. International journal of Research in Marketing, Volume 25, Issue 4, December 2008.
Barkow, J. H., **Cosmides**, L. and **Tooby**, J. (Editors), The Adapted Mind: Evolutionary Psychology and the Generation of Culture. Oxford University Press, 1992.
Bechara, A., **Damasio**, H. and **Damasio**, A. R., Emotion, decision-making and the orbitofrontal cortex. Cerebral Cortex, 10(3):295–307, 2000.
Beck, U., Risikogesellschaft. SuhrkampVerlag, Frankfurt am Main, 1986.
Bernstein, P. L., Against the Gods. Wiley, 1998.
Boltanski, L. and **Thévenot**, L., De la justification. Les économies de la grandeur. Gallimard, Paris, 1989.
Boyd, B., On the Origin of Stories: Evolution, Cognition and Fiction. Harvard University Press, 2009.
Boyer, P., Religion Explained. Basic Books, 2001.
Braun, C. C. and **Silver**, N. C., Interaction of signal word and colour on warning labels: differences in perceived hazard and behavioral compliance. Ergonomics, 38, 2207–2220, 1995.
Bryson, A. E. and **Denham**, W. F., A steepest-ascent method for solving optimum programming problems. Trans. of ASME, J. of Applied Mechanics, June, 1962.
Bryson Jr., A. E., Optimal control—1950 to 1985. Control Systems, IEEE, vol. 16, issue 3, June 1996.

Buss, D., The Evolution of Desire: Strategies of Human Mating. Basic Books, 2003.
Buss, D., Evolutionary Psychology: The New Science of the Mind. Pearson new international edition, fourth edition, 2014.
Carr, E. W., **Korb**, S. **Niedenthal**, P. M. and **Winkielman**, P. The two sides of spontaneity: Movement onset asymmetries in facial expressions influence social judgments. Journal of Experimental Social Psychology. 55, 31–36, 2014.
Carrère, S., **Buehlman**, K. T., **Gottman**, J. M., **Coan**, J. A. and **Ruckstuhl**, L., Predicting marital stability and divorce in newlywed couples. Journal of Family Psychology, Vol 14(1), 42–58, 2000.
Casscells, W., **Schoenberger**, A. and **Graboys**, T. B., Interpretation by physicians of clinical laboratory results. New England Journal of Medicine, 299(18):999–1001, 1978.
Chernoff, H., The Use of Faces to Represent Points in K-Dimensional Space Graphically. Journal of the American Statistical Association, 68 (342): 361–368, 1973.
Cialdini, R. B., Influence: The Psychology of Persuasion. Collins, 1984.
Clotfelter, C. T. and **Cook**, P. J., The "Gambler's Fallacy" in Lottery Play. Management Sciences, 39:12, 1521-1525, December 1993.
Combs, B. and **Slovic**, P., Causes of death: Biased newspaper coverage and biased judgments. Journalism Quarterly, 56: 837–843, 1979.
Connolly, T., Decision Theory, Reasonable Doubt, and the Utility of Erroneous Acquittals. Law and Human Behaviour, Vol. 11, no. 2. 1987.
Connolly, T., **Arkes** H. R. and **Hammond** K. R. (Editors), Judgment and Decision Making: An Interdisciplinary Reader. Cambridge University Press, 2000.
Cosmides, L. and **Tooby**, J., Cognitive Adaptations for Social Exchange. pp. 163–228, in Barkow, J. H., Cosmides, L. and Tooby, J. (Editors), 1992.
Cosmides, L. and **Tooby**, J., Are humans good intuitive statisticians after all? Rethinking some conclusions of the literature on judgment under uncertainty. Cognition, 58:1–73, 1996.
Cosmides, L. and **Tooby**, J., Unraveling the enigma of human intelligence: Evolutionary psychology and the multimodular mind. pp. 145–198, in R. J. Sternberg and J. C. Kaufman (Eds.), The evolution of intelligence, Hillsdale, NJ: Erlbaum, 2001.
Croson R. and **Sundali** J., The Gambler's Fallacy and the Hot Hand: Empirical Data from Casinos. The Journal of Risk and Uncertainty, 30:3; 195–209, 2005.
Darwin, C., The Origin of Species. Signet Classics, 2010 (orig. 1859).
Darwin, C., The expression of the emotions in man and animals. Harper Collins, 1998 (orig. 1872).
Davies, S., **Haines**, H., **Norris**, B. and **Wilson**, J. R., Safety pictograms: are they getting the message across?. Applied Ergonomics, Vol. 29, no. 1, 15–23. 1998.
Dawes, R. M., Proper and Improper Linear Models. In Connolly, T., Arkes, H. R. and Hammond, K. R. (Editors), 2000.
Dawes, R. M., The Robust Beauty of Improper Linear Models in Decision Making. American Psychologists, vol. 34, No 7, pp. 571–582, July 1979.
Dawkins, R., The Selfish Gene. Oxford University Press, 1976.

Dawkins, R. and **Krebs**, J. R., Arms races between and within species. Proceedings of the Royal Society, London Biological Society, 205(1161), 489-511, 1979.

Dehaene, S., Les neurones de la lecture. Odile Jacob, Paris, 2007.

Dellavigna, S., Psychology and economics: Evidence from the field. Journal of Economic Literature, 47:315–372, June 2009.

Denes-Raj., V. and **Epstein**, S., Conflict between intuitive and rational processing: When people behave against their better judgment. Journal of Personality and Social Psychology, 66(5):819–829, 1994.

Dennett, D. C., Consciousness Explained. Penguin Books, 1991.

Do A. M., **Rupert** A. V. and **Wolford** G., Evaluations of pleasurable experiences: the peak-end rule. Psychon. Bull. Rev. 15(1):96–8, 2008.

Dohmen, T., **Falk**, A., **Hufman**, D., **Marklein**, F. and **Sunde**, U., Biased probability judgment: Evidence of incidence and relationship to economic outcomes from a representative sample. Journal of Economic Behavior and Organization, 72(3):903–915, 2009.

Dunbar, R. I. M., Coevolution of neocortical size, group size and language in humans. Behavioral and Brain Sciences, 16 (4): 681–735, 1993.

Dunning, D., **Van Boven**, L. and **Loewenstein**, G. F., Egocentric empathy gaps in social interaction and exchange. In (ed.) Advances in Group Processes (Advances in Group Processes, Volume 18) Emerald Group Publishing Limited, pp.65–97, 2001

Dupuy, J. P., Pour un catastrophisme éclairé. Seuil, Paris, 2000.

Etcoff, N., Survival of the Prettiest: The Science of Beauty. Anchor Books, 2000.

Epstein, R. A., The Theory of Gambling and Statistical Logic. Academic Press, 1977.

Feller, W., An introduction to Probability Theory and Its Applications, Volume 1. W. Wiley, New York, 1968.

Fischhoff, B., **Slovic**, P., **Lichtenstein**, S., **Read**, S. and **Combs**, B., How safe is safe enough? A psychometric study of attitudes towards technological risks and benefits. Policy Sciences, 9(2):127–152, 1978.

Fischhoff, B., Quality and Safety in Health Care. British Medical Journal 12(4):304–311, August 2003.

Fischhoff, B., **Watson**, S. R. and **Hope**, C., Defining risk. Policy Sciences, 17(2):123139, October 1984.

Forer, B. R., The fallacy of personal validation: a classroom demonstration of gullibility. Journal of Abnormal and Social Psychology, 44, 118–123, 1949.

Frazer, J. G., The Golden Bough. Wordsworth editions Ltd, 1993 (orig. 1890).

Frederick, S.W., Automated Choice Heuristics. pp. 548-558, in Gilovich, T., Griffin, D. and Kahneman, D. (Editors), 2002.

Frederickson, B. L. and **Kahneman** D., Duration neglect in retrospective evaluations of affective episodes. Journal of Personality and Social Psychology, 65 (1): 45–55, 1993.

Galperin, A., **Fessler**, D. M. T., **Johnson**, K. L. and **Haselton**, M. G., Seeing storms behind the clouds: Biases in the attribution of anger. Evolution and Human Behavior, Volume 34, Issue 5, 358–365, 2013.

Garcia Retamero, R. and **Dhami**, M. K., Take the best in expert novice decision strategies for residential burglary. Psychonomic Bulletin & Review, 16(1), 163–169, 2009.

Gazzaniga, M. S., Natures's Mind: The Biological Roots of Thinking, Emotions, Sexuality, Language, and Intelligence. Perseus Group, 1992.
Gigerenzer, G., I think therefore I err. Social Research, 72(1): 195–218, 2005.
Gigerenzer, G., Rationality for Mortals. Oxford University Press, 2008.
Gigerenzer, G. and **Hoffrage**, U., How to improve Bayesian reasoning without instruction: Frequency Formats. Psychological Review, Vol. 102, No 4, 684–704, 1995.
Gilovich, T. and **Griffin**, D. and **Kahneman**. D. (Editors), Heuristics and biases: The psychology of intuitive judgement. Cambridge University Press, 2002.
Goldacre, B., Bad Science. Harper Perennial, 2009.
Goldstein, D. G. and **Gigerenzer**, G., Models of ecological rationality: The recognition heuristics. Psychological Review, 109(1):75–90, 2002.
Goldstein, N. J., **Martin**, S. J. and **Cialdini**, R., Yes! 50 Scientifically Proven Ways to Be Persuasive. Profile Books, Free Press, 2009.
Grice, H. P., Logic and Conversation, Syntax and Semantics, vol. 3 edited by P. Cole and J. Morgan. Academic Press, 1975.
Griskevicius, V. and **Kenrick**, D. T., Fundamental motives: How evolutionary needs influence consumer behavior. Journal of Consumer Psychology, Volume 23, Issue 3, 372–386, 2013.
Haidt, J., **McCauley**, C. and **Rozin**, P., Individual differences in sensitivity to disgust: a scale sampling seven domains of disgust elicitors. Personality and Individual Differences, 16, 701–713, 1994.
Harris, J., The Nurture Assumption: Why Children Turn Out the Way They Do. The Free Press, 2009.
Hartmann, P. and **Apaolaza-Ibáñez**, V., Beyond savanna: An evolutionary and environmental psychology approach to behavioral effects of nature scenery in green advertising. Journal of Environmental Psychology, Volume 30, Issue 1, 119–128, 2010.
Haselton, M. G. and **Nettle**, D., The paranoid optimist: An integrative evolutionary model of cognitive biases. Personality and Social Psychology Review, 10(1):47–66, 2006.
Healey, P. and **Glanvill**, R., Now That's What I Call Urban Myths. Virgin, 1996.
Henslin, J. M., Craps and magic. American Journal of Sociology, 73, 316–330, 1967.
Hill, R. A. and **Dunbar**, R. I. M., Social network size in humans. Human Nature, 14:53–72, 2003.
Hsee, C. K., Less Is Better: When Low-value Options Are Valued More Highly than High-value Options. Journal of Behavioral Decision Making, 11: 107–121, 1998.
Huber, V. L., **Neale**, M. A. and **Northcraft**, G. B., Decision bias and personnel selection strategies. Organ. Behav. Hum. Decis. Process, 40:136-147, 1987.
Hutchinson, J. M. C. and **Gigerenzer**, G., Simple heuristics and rules of thumb: Where psychologists and behavioural biologists might meet. Behavioural Processes. 69:2, 97–124, 2005.
Hyman, R., Cold Reading: How to Convince Strangers That You Know All About Them. Skeptical Inquirer, 1977.
Jackson, R. E. and **Cormack**, L. K., Evolved navigation theory and the descent illusion. Perception & Psychophysics, 69(3):353–62, 2007.

Jennings, D. L., **Lepper**, M. R. and **Ross**, L., Persistence of impressions of personal persuasiveness: Perseverance of erroneous self-assessments outside the debriefing paradigm. Personality and Social Psychology Bulletin, 7, 257–263, 1981.

Johnson, E. J. and **Tversky**, A., Affect, generalization, and the perception of risk. Journal of Personality and Social Psychology, 20–31, 1983.

Joffe, H., The power of visual material: persuasion, emotion and identification. Diogenes, 55(1):84–93, 2008.

Joffe. H. and **Haarhoff**, G., Representations of far-flung illnesses: the case of Ebola in Britain. Social Science and Medicine, 54(6):955–969, 2002.

Jonason, P. K., **Lyons**, M. and **Bethell**, E., The making of Darth Vader: Parent–child care and the Dark Triad. Personality and Individual Differences, 67, 30–34, 2014.

Jones, E .E. and **Harris**, V. A., The attribution of attitudes. Journal of Experimental Social Psychology, 3:1–24, 1967.

Kahneman, D. and **Miller**, D., Norm theory: Comparing reality to its alternatives. Psychology Review, 93(2):136–153, 1986.

Kahneman D. and **Thaler** R. H., Utility Maximization and Experienced Utility. Journal of Economic Perspectives, 2006.

Kahneman, D. and **Tversky**, A., Judgment under uncertainty: Heuristics and biases. Science, 185, 1124–1131, 1974.

Kahneman, D. and **Tversky**, A., Prospect theory: An analysis of decision under risk. Econometrica, 47(2):263–292, 1979.

Kahneman, D. and **Tversky**, A., Intuitive prediction: Biases and corrective procedures. pp. 414–421 in Kahneman, D., Slovic, P. and Tversky, A. (Editors), 1982.

Kahneman, D., **Slovic**, P. and **Tversky**, A. (Editors), Judgment Under Uncertainty: Heuristics and Biases. Cambridge University Press, 1982.

Kahneman, D., **Knetsch**, J. L. and **Thaler**, R. H., The endowment effect, loss aversion, and status quo bias: Anomalies. Journal of Economic Perspectives, 5(1):193–206, Winter 1991.

Kapferer, J.-N. Rumours: Uses, Interpretations and Images. Transaction Publishers, New Brunswick 1990.

Kramer, R. M., The sinister attribution error: Paranoid cognition and collective distrust in organizations. Motivation and Emotion, 18(2):199–230, 1994.

Kunreuther, H. and **Slovic**, P., Economics, psychology, and protective behavior. The American Economic Review, 68(2): 64–69, 1978.

Kurz-Milcke, E., **Gigerenzer**, G. and **Martignon**, L., Transparency in Risk Communication. Annals of the New York Academy of Sciences, 1128: 18–28, 2008.

Lakoff, G. and **Johnson**, M., Metaphors We Live By. University of Chicago Press, 1980.

Lakoff, G. and **Johnson**, M., Philosophy in the Flesh: the Embodied Mind and Its Challenge to Western Thought. New York, Basic Books, 1999.

Langer, E., The Illusion of Control. Journal of Personality and Social Psychology, 32 (2): 311–328, 1975.

Leakey, R. E. and **Lewin**, R., The Origins of Man., E. P. Dutton, New York, 1977.

LeDoux, J., The Emotional Brain. Phoenix Division of Orion Book, 1998.

Lesch, M. F., **Rau**, P. P., **Zhao**, Z. and **Liu**, C., A cross-cultural comparison of perceived hazard in response to warning components and configurations: US vs. China. Applied Ergonomics, 40(5):953–961, 2009.

Levenson, M. R., Risk taking and personality. Journal of Personality and Social Psychology, 58(6):1073–1080, 1990.

Levin, I. P. and **Gaeth**, G. J., How Consumers are Affected by the Framing of Attribute Information Before and After Consuming the Product. Journal of Consumer Research, 15, 374–378, 1988.

Loewenstein, G. F. and **Thaler**, R. H., Anomalies: Intertemporal choice. Journal of Economic Perspectives, 3(4):181–193, Autumn 1989.

Loewenstein, G. F., **Weber**, E. U., **Hsee**, C. K. and **Welch**, N., Risk as feelings. Psychological Bulletin, 127(2):267–286, 2001.

Marks, I. M. and **Nesse**, R. M., Fear and fitness: An evolutionary analysis of anxiety disorders. Ethology and Sociobiology, 15:247–261, 1994.

Martignon, L. and **Krauss**, S., Hands-on activities for fourth graders: a tool box for decision-making and reckoning with risk. International Electronic Journal of Mathematics Education, 4(3), October 2009.

McFadden, D., Free Markets and Fettered Consumers. American Economic Review, Vol. 96, No 1, 5–29, 2006.

Medvec, V. H., **Madey**, S. F. and **Gilovich**, T., When less is more: Counterfactual thinking and satisfaction among Olympic medalists. Journal of Personality and Social Psychology, 69:603–610, 1995.

Meehl, P., Clinical versus Statistical Prediction: A Theoretical Analysis and a Review of the Evidence. Echo Point Books and Media, 2013 (orig. 1954).

Meissner, C. A. and **Kassin**, S. M., "He's guilty!": Investigator bias in judgments of truth and deception. Law and Human Behavior, 26(5), 469–480, Oct 2002.

Meyerowitz, B. E and **Chaiken**. S., The effect of message framing on breast self-examination attitudes, intentions and behaviour. Journal of Personality and Social Psychology 52(3):500–510, 1987.

Milgram, S., **Bickman** L. and **Berkowitz** L., Note on the drawing power of crowds of different size. Journal of Personality and Social Psychology, 13:79–82, 1969.

Miller, G. A., The Magical Number Seven, Plus or Minus Two: Some Limits on Our Capacity for Processing Information. Psychological Review, 101(2), 343–352, 1955.

Miller, D. and **Taylor**, B. R., Counterfactual Thought, Regret, and Superstition: How to Avoid kicking Yourself. In Gilovich, T. and Griffin, D. and Kahneman, D. (Editors), 2002.

Neuhoff, J. G., Perceptual bias for rising tones. Nature, 395 (6698), 123–124, 1998.

"Mystery", The Mystery Method: How to Get Beautiful Women into Bed. St. Martin's Press, New York, 2007.

Mumpower, J. L., Lottery Games and Risky Technologies: Communications about Low-probability/high-consequence events. Risk Analysis, 8:2, 231–5, 1988.

Nisbett, R. E. and **Ross**, L., Human inference: Strategies and shortcomings of social judgment. Prentice-Hall, Englewood Cliffs, NJ, 1980.

Nisbett, R. E., **Borgida** E., **Crandall** R. and **Reed** H., Popular Induction: Information is Not Always Informative. In J. S. Carroll and J. W. Payne (Editors), Cognition and Social Behavior, Halsted, 1976.

Olson, E. A. and **Wells** G. L., What makes a good alibi? A proposed taxonomy. Law and Human Behavior, 28(2):157–76, 2004.

Ong, A. D. and **Weis**, D. J., The Impact of Anonymity on Responses to Sensitive Questions. Journal of Applied Social Psychology, 30, 8,1691–1708, 2000.

Panagopoulos, C., Watchful eyes: implicit observability cues and voting. Evolution and Human Behavior, Volume 35, Issue 4, 279–284, 2014.

Paulhus, D. L., Measurement and control of response bias. In J. P. Robinson, P. R. Shaver & L. S. Wrightsman (Eds.), Measures of personality and social psychological attitudes (pp. 17–59). San Diego, CA: Academic Press, 1991.

Pinker, S., The Language Instinct: The New Science of Language and Mind. Penguin Science, 1995.

Pinker, S., The Blank State: The Modern Denial of Human Nature. Viking, 2012.

Pinker, S., How the Mind Works, W. W. Norton, 1997.

Popper, K., Conjectures and Refutations. Routledge Classics, 2002.

Quian Quiroga, R., **Reddy**, L., **Kreiman**, G., **Koch**, C. **and Fried**, I. Invariant visual representation by single neurons in the human brain. Nature, 435 (7045), pp. 1102-1107, 2005.

Rabin, M. and **Schrag**, J., First impressions matter: A model of confirmatory bias. Quarterly Journal of Economics, 114(1):37–82, 1999.

Read, D., **Loewenstein**, G. and **Kalyanaraman**, S., Mixing virtue and vice: Combining the immediacy effect and the diversification heuristic. Journal of Behavioral Decision-making, 12, 257–273, 1999.

Redelmeier, D. A. and **Kahneman** D., Patients' memories of painful medical treatments: real-time and retrospective evaluations of two minimally invasive procedures. Pain 66(1):3–8, 1996.

Richard-Zappella, J., Variations interrogatives dans la question de sondage. Mots, n°23, juin 1990.

Rind, B., Effect of beliefs about weather conditions on tipping. Journal of Applied Social Psychology, 26, 137–147, 1996.

Rozin, P. and **Fallon**, A., A perspective on disgust. Psychological Review, 94(1):23–41, 1987.

Samuelson, P. A., Risk and uncertainty: A fallacy of large numbers. Scientia, April-May, 1963.

Samuelson, W. and **Zeckhauser**, R., Status quo bias in decision-making. Journal of Risk and Uncertainty, 1(1):7–59, March 1988.

Sandman, P. M., **Miller**, P. M., **Johnson**, B. B. and **Weinstein**, N. D., Agency communication, community outrage, and perception of risk: Three simulation experiments. Risk analysis, 13(6):585–598, 1993.

Saunders, E. M., Stock prices and Wall Street weather. American Economic Review, 83(5):1337–45, December 1993.

Schott, J. L., No Left Turns: The FBI in Peace and War. New York, Prager, 1975.

Schwartz, B., **Ward**, A., **Monterosso**, J., **Lyubomirsky**, S., **White**, K. and **Lehman**, D., Maximising versus Satisficing: Happiness is a Matter of Choice. Journal of Personality and Social Psychology, Vol. 2, No. 5, 1178–1197, 2002.

Shafir, E., Choosing versus rejecting: Why some options are both better and worse than others. Memory & Cognition, 21, 4, 546–556, 1993.

Shepard, N., The Perceptual Organization of Colors: An Adaptation to Regularities of the Terrestrial World?. In Barkow, J. H., Cosmides, L., and Tooby, J., 1992.

Sim, T. C. and **Martinez**, C., Emotion words are remembered better in the left ear. Laterality Volume 10, Issue 2, 149–159, 2005.

Simon, H. A., Models of Bounded Rationality. MIT Press, Cambridge MA, 1982.

Sjöberg, L., Factors in risk perception. Risk analysis, 20(1): 1–11, 2000.

Skinner, B. F., 'Supersitition' in the pigeon. Journal of Experimental Psychology, 38, 168–172, 1948.

Slovic, P., **Fischhoff**, B. and **Lichtenstein**, S., Facts versus fears: Understanding perceived risk. pp. 463–490, in Kahneman, D., Slovic, P. and Tversky, A. (Editors), 1982.

Slovic, P., **Finucane**, M. L., **Peters** E. and **MacGregor**, D. G., The Affect Heuristic. pp. 397–420, in Gilovich, T. and Griffin, D. and Kahneman, D.(Editors), 2002.

Slovic, P., **Finucane**, M. L., **Peters** E. and **MacGregor**, D. G., Risk as analysis and risk as feelings: Some thoughts about affect, reason, risk, and rationality. Risk Analysis, 2(24):311–322, 2004.

Sobell, L. C., Bridging the gap between scientists and practitioners: The challenge before us. Behavior Therapy, Volume 27, Issue 3, Pages 297–320, Summer 1996.

Stewart, A., On risk: perception and direction. Computer and Security, 23:362–370, 2004.

Strack, F. and **Mussweiler**, T., Explaining the enigmatic anchoring effect: Mechanisms of selective accessibility. Journal of Personality and Social Psychology, 73, 437–446, 1997.

Sutherland, C. A. M., **Thut**, G. and **Romei**, V., Hearing brighter: Changing in-depth visual perception through looming sounds. Cognition 132, p. 312–323, 2014.

Svenson, O., Are we less risky and more skillful than our fellow drivers? ActaPsychologica, 47, 1981 143–151, 1981.

Swets, J. A, Enhancing Diagnostic Decisions. pp. 66–81, in Connolly, T., Arkes, H. R. and Hammond, K. R. (Editors), 2000.

Taleb, N. N., Fooled by Randomness. Penguin Books, 2004.

Taleb, N. N., The Black Swan. Penguin Books, second edition, 2010.

Thaler, R. H., Toward a positive theory of consumer choice. Journal of Economic Behavior and Organization, 1:39–60, 1980.

Thaler, R. H., Mental accounting and consumer choice. Marketing Science, 4, 199–214, 1985.

Thaler, R. H., Illusions and Mirages in Public Policy. pp. 85–96, in Connolly, T., Arkes, H. R. and Hammond, K. R. (Editors), 2000.

Thaler, R. H. and **Sunstein**, C. R., Nudge: Improving Decisions about Health, Wealth, and Happiness. Yale University Press, 2008.

Tierney, J., Behind Monty Hall's Doors: Puzzle, Debate and Answer?. New York Times, July 21, 1991.
Trivers, R. L., Parental investment and sexual selection. pp. 136–179, in B. Campbell (Ed.), Sexual selection and the descent of man, 1871–1971, Chicago, Aldine, 1972.
Tversky, A. and **Kahneman**, D., Rational choice and the framing of decisions. Journal of Business, Vol. 59, No. 2, 251–278, 1986.
Wason, P. C., On the failure to eliminate hypotheses in a conceptual task. Quarterly Journal of Experimental Psychology, 12:3,129–140, 1960.
Wason, P. C., Reasoning about a rule. Quarterly Journal of Experimental Psychology, 20(3):273-281, 1968.
Weber, E. U., **Hsee**, C. K. and **Sokolowska**, J., What folklore tells us about risk and risk taking: Cross-Cultural Comparisons of American, German, and Chinese Proverbs. Organizational behavior and human decision processes, Vol. 75, No. 2, 170–186, 1998.
Wedell, D. H., **Santoyo**, E. M. and **Pettibone**, J. C., The Thick and the Thin of It: Contextual Effects in Body Perception. Basic and Applied Social Psychology, 27(3), 213–227, 2005.
Weinstein, N. D., Unrealistic optimism about future life events. Journal of Personality and Social Psychology, 39: 806–820, 1980.
Wells, G. L. and **Petty**, R., The effects of head movement on persuasion: Compatibility and incompatibility of responses. Basic and Applied Social Psychology, 1, 219–230, 1980.
Willems, R. M., **Hagoort**, P. and **Casasanto**, D., Body-specific representations of action verbs: neural evidence from right- and left-handers. Psychological Science, 21(1):67–74, 2010.
Wilson, E. O., Sociobiology: The New Synthesis. Twenty-Fifth Anniversary Edition, Harvard Press, 2000.
Wiseman, R., 59 Seconds, Pan Books, 2010.
Wogalter, M. S. and **Mayhorn**, C. B., Providing cognitive support with technology-based warning systems. Ergonomics, 48(5):522–533, April 2005.
Wogalter, M. S., **Kalsher**, M. J. and **Rashid** R., Effect of signal word and source attribution on judgments of warning credibility and compliance likelihood. International Journal of Industrial Ergonomics, 24, 185–192, 1999.
Yu, R. F., **Chan**, A. H. S. and **Salvendy**, G., Chinese perceptions of implied hazard for signal words and surround shapes. Human Factors and Ergonomics in Manufacturing & Service Industries, 14, 1, 69–80, 2004.
Zajonc, R. B., Feeling and thinking: Preferences need no inferences. American Psychologist, 35(2):151–175, February 1980.
Zajonc, R. B., **Adelmann**, P. K., **Murphy**, S. T. and **Niedenthal**, P. M., Convergence in the physical appearance of spouses. Motivation and Emotion, 11:335–346, 1987.
Zeckhauser, R. J. and **Viscusi**, W. K., Risk within Reason pp. 465–478, in Connolly, T., Arkes, H. R. and Hammond, K. R. (Editors), 2000.

Printed in the United States
By Bookmasters